T0325139

Airborne Electronic Hardware Design Assurance

Airborne Electronic Hardware Design Assurance

A Practitioner's Guide to RTCA/DO-254

Randall Fulton

SOFTWAIR ASSURANCE, INC.
REDWOOD CITY, CA

Roy Vandermolen

HARDWAIR ASSURANCE
BOUNTIFUL, UT

CRC Press is an imprint of the
Taylor & Francis Group, an **informa** business

CRC Press
Taylor & Francis Group
6000 Broken Sound Parkway NW, Suite 300
Boca Raton, FL 33487-2742

International Standard Book Number-13: 978-1-4822-0605-0 (Hardback)

Library of Congress Cataloging-in-Publication Data

Fulton, Randall.
Airborne electronic hardware design assurance : a practitioner's guide to RTCA/DO-254 / Randall Fulton, Roy Vandermolen.
pages cm
Includes bibliographical references and index.
ISBN 978-1-4822-0605-0 (hardback)
1. Avionics--Standards--United States. 2. Guidance systems (Flight)--Standards--United States. I. Vandermolen, Roy. II. RTCA (Firm) III. Title.

TL695.F85 2014
629.13502'1873--dc23 2014043747

Visit the Taylor & Francis Web site at
http://www.taylorandfrancis.com

and the CRC Press Web site at
http://www.crcpress.com

Contents

Preface

This book was inspired by, and evolved from, the experience we gained while teaching the DO-254 class at RTCA in Washington, D.C. and at numerous companies throughout the United States and abroad. In some respects it is the logical extension of our DO-254 class, and in fact some of the material in this book shares common origins.

Most of this book, however, is a testimony of the lessons learned and wisdom gained from many years of first-hand experience in the design, verification, and approval of airborne electronics and software, including some of the very first projects in which DO-254 was applied, and also including the very first projects where DO-254 was applied at the LRU level. The early years of DO-254 were seminal to us for revealing not only the difficulties inherent in applying a comprehensive document like DO-254 to a very narrow target, but also for the labor pains endemic throughout the industry as it struggled to understand—let alone comply with—this new way of doing business. Pain is a hard but effective teacher, so by any measure those of us who experienced those years should be geniuses by now.

As we and the rest of the industry adjusted and eventually mastered both the document and its ramifications, we were able to discern the practices and techniques that complemented the processes in DO-254 and therefore worked best in this new environment. This book documents the practices and techniques that we have identified and witnessed as being compatible with the intent of DO-254, and which thus make the road to compliance and eventual approval as direct, efficient, and effective as possible. While most of the material in this book is written from the perspective of programmable logic devices (à la FAA Advisory Circular 20-152), all of it can easily be extended to encompass the rest of an electronic system in the way DO-254 was actually intended to be used.

That said, there is only so much that can be documented, let alone taught, in one book. The concepts and techniques that we introduce here are presented in their most basic and fundamental form; actually mastering the topic can only be accomplished through first-hand experience in developing an electronic system in a certification program. While this means that newcomers to the world of DO-254 compliance may experience the sometimes difficult initiation that was sustained by those who went before them, it is our fond desire that they will embrace the material in this book to minimize the cost—both literal and metaphorical—to its lowest possible level. After all, a wise person will learn from his own mistakes, but a wiser person will learn from other peoples' mistakes. This book is built upon the mistakes and successes of many people all over the world, so it offers the wiser person the means to learn from the mistakes of a distinguished club of professionals who had to learn their lessons through experience, hardship, and even trial and error. Through this book the reader can access an enormous compendium of experiences without having to experience the associated travail first hand.

It is our desire, and our motivation for writing this book, to make that wisdom available to all of the developers of safety critical electronic hardware to minimize the needless repetition of the difficulty of those early years, and of course to make modern aircraft as safe and reliable as possible.

Acknowledgments

This book is dedicated to all the engineers, managers, and personnel we encounter in the trainings, consulting, development programs, and compliance programs that we have and will participate in. We would like to express our utmost gratitude to all those people who have taken the time to help us further our understanding and application of design assurance in the realm of safety critical system and electronics development.

We also thank our partners, friends, and colleagues for their support in this effort.

Authors

Randall Fulton is an FAA Consultant DER with over 36 years of Electrical Engineering experience emphasizing software and electronic hardware development and verification. Randall has a Bachelor of Science in Electrical Engineering from the Pennsylvania State University and earned his Federal Aviation Administration Designated Engineering Representative credentials in software and programmable logic devices while working at Boeing Commercial Aircraft. As a Designated Engineering Representative, Randall has had approval authority for programmable logic devices since 1997 and has worked numerous Part 23 and Part 25 certification programs with field programmable gate arrays, application-specific integrated circuits (ASIC), and software. Randall, along with Roy Vandermolen, taught the DO-254 practitioners course for RTCA in Washington, D.C. from 2006 to 2009. Randall has also taught the Airborne Electronic Hardware Job Functions class for the Federal Aviation Administration Academy in Oklahoma City. Randall works as a Consultant DER through his company SoftwAir Assurance, Inc.

Roy Vandermolen is an electronics design engineer with over 35 years of experience in electronics ranging from vacuum tubes to programmable logic devices, but has spent the majority of that time designing and verifying programmable logic devices and the circuit cards that employ them. Roy has a Bachelor of Science in Electrical Engineering from the Massachusetts Institute of Technology (MIT), and is currently a Staff Engineer and Certification Manager for electronic flight control systems at Moog Aircraft, and an airborne electronics hardware Outside Boeing Authorized Representative (OBAR) for Boeing Commercial Aircraft. Roy has worked in a variety of engineering environments including research laboratories, educational institutions, military R&D facilities, and commercial aircraft flight controls manufacturers. In his career at Moog Aircraft Roy has been involved in the design, verification, and certification of numerous Level A flight control systems. Roy, along with Randall Fulton, taught the DO-254 practitioners course for RTCA in Washington, D.C. from 2006 to 2009.

Authors

1 Introduction to RTCA/DO-254

RTCA DO-254 (also known as EUROCAE ED-80), *Design Assurance Guidance for Airborne Electronic Hardware*,[1] was prepared jointly by RTCA Special Committee 180 and EUROCAE Working Group 46, and was subsequently published by RTCA and EUROCAE in April 2000. These industry working groups were formed in the 1990s and took seven years to create DO-254.

DO-254 is not a technical manual, nor is it an engineering cookbook. It does not prescribe the design characteristics of electronic circuits or components, nor does it contain design standards. In fact, it contains virtually no technical information that an engineer can use to guide a circuit design. Instead, it provides guidance on the processes and methodologies that should be applied in the development and verification of electronic hardware to achieve an acceptable level of confidence that the end hardware functions correctly and will be in compliance with airworthiness requirements. While it can be argued that some aspects of a design can be influenced by the desire to facilitate the design and verification methodologies contained in DO-254, there are actually no design features that must or must not exist solely because of the need to comply with it.

The guidance in DO-254 represents industry consensus on the best practices that will ensure that electronic hardware is developed and verified in a way that is appropriate for its design assurance level (DAL)—often shortened to "Level"—and will ensure, to a realistic level, a safe and reliable product. The supporting processes defined in DO-254 (configuration management, process assurance, and validation/verification) are particularly effective in controlling the introduction of design errors as well as identifying errors that are inevitably introduced despite best efforts to the contrary. DO-254 describes the objectives for each phase of a typical electronic hardware development life cycle and describes the activities usually associated with the life cycle phase.

The genesis of DO-254 stems from concerns that as electronics technology rapidly evolved, enabling systems to become more complex and to host more functionality, proving that these systems were safe and reliable was becoming more and more difficult. Most of the electronic systems and programmable logic devices (PLDs) that are used on modern aircraft are well beyond our ability to prove safe through quantitative analysis, and in the absence of this avenue of design assurance, the only other viable means of establishing the necessary design assurance is to use structured and disciplined processes and methodologies during their development. The advantages of using this means are, first, that the processes and methodologies force designers to create a design in a logical and systematic (and therefore repeatable) manner, and second, they create a type of transparency in the design and its project by enforcing a high level of documentation and traceability that, while inconvenient on the surface, has repeatedly proven its worth over the long-term life of a design. Like them or not, the best practices in DO-254 can, particularly in the long run, be a project's best friend.

The guidance in DO-254 was written to apply to all complex electronic hardware that performs safety-critical system functions. While DO-254 discusses PLDs, they are considered within the context of system and equipment development, and not necessarily as the only aspect of the system that should use the guidance in DO-254. When the Federal Aviation Administration (FAA) published its Advisory Circular (AC) 20-152,[2] which approved DO-254 as an acceptable (but not the only) means of satisfying the Federal Aviation Regulations (FARs) for PLDs, it essentially reduced the application of DO-254's guidance to a small subset of its original scope.

Narrowing the focus of DO-254 from the system level to the component level can be problematic due to the need to interpret and apply electronic system guidance to singular components within the system. Examples of potential problems include determining the scope of the life cycle data, determining how to interpret and apply DO-254 Table A-1 to PLD life cycle data, and sorting out which activities and aspects apply at the component level as opposed to the system level (such as acceptance tests, environmental tests, functional failure path analysis, and traceability to elements in the hardware implementation, none of which are entirely practical at the component level). There are also boundary issues that arise: much of DO-254's overall effectiveness relies upon implementing all of its guidance to all levels of a system, which creates a seamless interconnection and flow of processes and data between all levels of the system (line replaceable unit [LRU], sub-assemblies, circuit card assemblies [CCA], and component [typically a PLD]). When one level of the system is singled out for the isolated application of DO-254, it loses this flow to the other levels, creating a discontinuity in both processes and data that can, if not properly anticipated, potentially render much of the design assurance activities ineffective or meaningless. Applying DO-254 only at the PLD level can be made to work through judicious interpretation and tailoring, but in the end the document is most easily comprehended and most effectively applied according to the perspective for which it was conceived and written, and ultimately for which it was intended to be applied, in other words throughout the entire system.

DESIGN ASSURANCE LEVEL

DO-254 defines five levels for the design assurance of airborne electronic systems. These five design assurance levels are defined as levels A through E, where A is the most stringent and E is the least. These five levels correspond to the five classifications of failure conditions defined in the regulatory materials that govern the certification of airborne systems and equipment.

Table 1.1 identifies the five hazard classifications and maps them to their corresponding design assurance level, the required probability of failure per flight hour for equipment of each level, and a description of the hazard. The hazard classifications are described with respect to the effect that a failure of the system or equipment will have on the aircraft, its occupants, its safety margins, and the ability of

TABLE 1.1
Design Assurance Level and Hazard Classification

Failure Classification	Hardware Design Assurance Level	Probability of Failure per Flight Hour	Hazard Description
Catastrophic	Level A	10^{-9}	Prevents continued safe flight and landing.
Hazardous/ Severe-Major	Level B	10^{-7}	• Serious or fatal injuries to small number of occupants. • Reduces aircraft capabilities or crew's ability to deal with adverse operating conditions. • Higher crew workload. • Large reduction in safety margins.
Major	Level C	10^{-5}	• Possible injuries to occupants. • Reduces aircraft capabilities or crew's ability to deal with adverse operating conditions. • Increase in crew workload. • Significant reduction in safety margins.
Minor	Level D	10^{-3}	• Possible inconvenience to occupants. • Reduces aircraft capabilities or crew's ability to deal with adverse operating conditions. • Slight increase in crew workload. • Slight reduction in safety margins.
None	Level E	N/A	• No effect on operational capabilities. • No crew workload impact.

its crew to deal with adverse operating conditions. The most severe classification is *catastrophic* (level A), indicating that a failure of level A equipment will, for all practical purposes, result in a catastrophic hull loss of the aircraft. The least severe is *no effect*, indicating that a failure will affect neither the aircraft's operational capabilities nor the workload of its crew.

In DO-254 the design assurance level also determines the objectives of, and amount of rigor in, the development process, the type and quantities of artifacts from the development effort that must be preserved, the level of independence that must be maintained when conducting the development activities, and the amount and type of verification that must be performed on the design.

The guidance in the main body of DO-254 (chapters 1 through 11) is written for electronic hardware of design assurance level C. DO-254 accommodates the higher design assurance levels (A and B) through the application of its Appendix B, which contains additional verification-related activities that are intended to provide the extra measure of assurance that is appropriate for equipment that requires per-hour failure probabilities less than 10^{-7}. Equipment with a design assurance level of C or D must comply with chapters 1 through 11 of DO-254, and levels A and B must comply with chapters 1 through 11 and Appendix B.

It is worth noting that the inherent failure probability of semiconductor components is typically no better than 10^{-5} to 10^{-6}. Or in other words, the inherent reliability limitations of semiconductor devices prevents the electronic hardware from achieving the failure probabilities required for level A or B systems, and therefore the development guidance in DO-254 will not by itself result in a design with the requisite level A or B reliability. The requisite reliability, however, can be achieved at the system level through the architectural mitigation techniques described in SAE ARP4754[3] or SAE ARP4754A,[4] even though the hardware itself is limited to the approximate failure rate for level C hardware.

Since the design process cannot create hardware that is able by itself to satisfy the requisite failure rate for level A or B systems, the only remaining avenue of design assurance for those higher levels is to employ additional verification-related techniques to ensure that every design error is detected and removed, thereby assuring that the hardware will achieve its maximum potential reliability. DO-254 Appendix B, which focuses on the verification-related techniques that ensure that every function in the design is fully identified and tested, is applied to level A and B hardware to ensure, as well as possible, that all design errors are detected and removed.

DO-254 AND DO-178B

DO-178B,[5] *Software Considerations in Airborne Systems and Equipment Certification*, was published and adopted years before electronic hardware was required to comply with DO-254. DO-178B's head start had two rather curious effects on electronic hardware development. First, there was a temptation for equipment manufacturers to move their system functionality from software to hardware because of the significant increase in time and effort (both perceived and actual) that DO-178B imposed on software development. The idea, of course, was that if equipment functionality could be moved to electronic hardware it would not have to be subjected to the expensive and difficult design assurance processes in DO-178B. The second effect is that the terminology, concepts, and processes from DO-178B became so well established in the airborne design assurance community that it still heavily influences how DO-254 is applied and implemented for hardware. While it may seem obvious that hardware and software are fundamentally different, and therefore should be treated differently for the purposes of design assurance, this seemingly elemental observation has at times been overwhelmed by the sheer inertia of DO-178B's influence.

So how do DO-254 and DO-178B (or DO-178C[6]) compare? And are the differences all that significant? Since both documents provide guidance on development processes, there are bound to be significant similarities in their content and philosophy. In fact, from the high level perspective they are quite similar in their approaches and fundamental concepts. The similarities can be summarized as follows:

- Their safety background and basis are the same.
- Both rely on process and design assurance.
- Both use life cycle phases to govern development.
- Both use the integral processes of process assurance (quality assurance for software), configuration management, and verification.
- Verification is requirements-based.
- Both include tool qualification.

So in a broad sense the two documents are very similar. However, when examining the details of their design assurance concepts and processes, a number of differences emerge, some of them dramatic enough that trying to apply DO-178 concepts to hardware development will not result in the desired design assurance. The differences between DO-254 and DO-178 (and to some extent between hardware and software) are embedded in the details, and are listed in **Table 1.2**.

TABLE 1.2
Differences between DO-254 and DO-178B/C

Topic	DO-254 Hardware	DO-178B/C Software
Environmental tests	Required	Not applicable
Part wear out	Need to consider	N/A
White box testing	Test points and simulation	Can perform with emulator and/or symbolic debugger
Implementation	Hardware and components	Machine code within hardware components
Debugging	Device pin level	Assembler level
Robustness testing	Optional or project dependent	Required
Object oriented considerations	Not applicable	Ada95, C++, Java
Objectives for compliance	Defined in sections 4.1, 5.1.1, 5.2.1, 5.3.1, 5.4.1, 5.5.1, 6.1.1, 6.2.1, 7.1, 8.1	Annex A Tables
Applicability	LRU, CCA, PLD, any complex electronic hardware	Software only
Tool qualification	Tool qualification not required for elemental analysis tools	Tool qualification required for structural coverage analysis tools
Independence	Broad brush stroke approach in Appendix A	Specific objectives depending on software level

continued

TABLE 1.2 (continued)
Differences between DO-254 and DO-178B/C

Topic	DO-254 Hardware	DO-178B/C Software
Simulation	May need second independent simulation; difficult to use same test procedures in simulations and in-circuit hardware tests	Easy to use the same test cases on simulation and on target hardware
Coverage analysis	Elemental analysis not specific about coverage criteria definition	Modified Condition/Decision Coverage defined for Level A Decision coverage defined for Level B Statement coverage defined for Level C
Verification	Test coverage of requirements, coverage of elements for Levels A and B	Test coverage of high-level and low-level requirements, structural coverage and test coverage of data and control coupling
Definition of derived requirement	Design decision; may or may not have a parent requirement	Glossary—almost the same as DO-254 Usage in Section 5—no parent requirement
Simple	Exhaustive input testing and reduced documentation allowed	Not applicable
Design approach	For PLDs, hardware description language describes how the physical hardware in the PLD will be configured For other electronic hardware, graphical entry such as schematic diagrams	Procedural language that describes a sequence of steps
Processing	For PLDs, parallel simultaneous implementation of hardware design language (HDL); functionality is implemented concurrently	Instructions execute in sequence
Validation	Derived requirements are validated to ensure they are correct and complete	Not in the scope of software—derived requirements must be justified but are not validated
Means of application	Written for Level C with additional measures for Levels A and B	Written for Level A with reductions in objectives, activities, and artifacts for Levels B through D

While some of the differences may seem trivial or inconsequential, others are significant and can result in serious consequences if DO-178 is allowed to influence how DO-254 is applied to hardware. All of the differences should be carefully considered and understood to pre-empt any confusion or misunderstandings when applying DO-254. In general, software definitions, techniques, and processes should not be used with hardware.

Perhaps the most insidious, confusing, and persistent of the differences is the definition of a derived requirement. DO-178B's long dominance of the design assurance field has resulted in a strong tendency, especially among people who have a software background or who are used to the language of DO-178B, to assume that the DO-254 definition is the same as the one found in DO-178B, or more commonly, that the DO-178B definition is universal and therefore applies to hardware as well as software. However, while the glossaries in both documents define a derived requirement as an additional requirement that results from the design process and which (may or) may not be directly traceable to a higher level requirement, Section 5.0 of DO-178B narrows the definition of a derived requirement to a requirement that does not directly trace to a higher level requirement. The definition in DO-178B Section 5.0 has superseded the glossary definition and has, again due to the long-time dominance of DO-178B, been treated as the more or less universal definition of a derived requirement. This is unfortunate from a number of perspectives, a significant one being that the processes (particularly the validation process) in DO-254 are designed to work with the DO-254 definition, and as will be described in more detail later, problems can arise if the DO-178B definition is indiscriminately used with the processes in DO-254.

OVERVIEW OF DO-254

The guidance in DO-254 covers a variety of topics including:

- Hardware standards
- Hardware design life cycle data
- Additional design assurance techniques for design assurance level (DAL) A and B functions
- Previously developed hardware
- Tool assessment and qualification
- Use of commercial off-the-shelf (COTS) components
- Product service experience
- Hardware safety assessment
- Design assurance strategy, including consideration for DAL A and B functions
- Planning process
- Requirements capture
- Conceptual design
- Detailed design
- Implementation
- Production transition
- Validation process
- Verification process including tests and reviews
- Configuration management
- Process assurance
- Certification liaison and proposed means of compliance

DO-254 ties together several key aspects of equipment development and compliance issues that would be otherwise difficult to achieve for complex and/or highly integrated devices and/or systems. These aspects are:

- Use of design assurance in lieu of quantitative analysis of failures
- Use of requirements to capture the aspects of aircraft functions performed by complex electronic hardware
- Use of review, analysis, and test to satisfy compliance to the FARs through verification

DO-254 contains eleven sections (or chapters) and four appendices. **Table 1.3** identifies the sections and appendices and summarizes their topic areas.

The sections of DO-254 and their relationships are shown in **Figure 1.1**.

Section 1 introduces DO-254 and describes its scope and applicability, and defines certain keywords such as "should," "may," and "hardware item." It is worth noting here that DO-254 does not define nor use the word "shall," which is commonly used to identify mandatory practices or requirements. This is appropriate given DO-254's role as *guidance* rather than requirements. Even AC 20-152 does not mandate compliance to DO-254 by stating that compliance to DO-254 is an acceptable, but not the only, means of complying with the FARs. Novices in the certification industry are quick to latch onto this point and object that DO-254 is just guidance, not requirements, and that the FAA does not make compliance to it mandatory. This is normally stated in the hope of sidestepping DO-254, but in realistic terms, any means of compliance other than DO-254 could be much more difficult and expensive to implement, even in the unlikely event that the certification authorities approve it.

DO-254 Section 1 also states that DO-254 does not attempt to define the nature or identity of "firmware," but does state that firmware should be classified as either hardware or software and then addressed appropriately through DO-254 or DO-178 as applicable. To persons who are inexperienced in the business of hardware description languages and PLD design, it is common for an HDL's resemblance to software to prompt them to try to classify it as such. However, those who are more experienced with firmware realize that HDLs, while similar to software code in appearance, are fundamentally different and have more in common with, and thus should be classified and treated as, hardware. Subjecting an HDL design to the software processes in DO-178B or DO-178C would present a host of problems, and for some processes would prove to be impractical if not impossible.

One aspect of DO-254 Section 1 that attracts a great deal of interest is the introduction of the concept of "simple" versus "complex" hardware, especially the last paragraph of Section 1.6, which states that simple hardware items do not need extensive documentation and can comply with DO-254 with reduced overhead. For virtually all equipment suppliers, the idea of simple hardware, and the reduced effort associated with it, is immensely attractive, and is immediately embraced as a possible means of avoiding full compliance to DO-254. This desire to avoid the cost and schedule impact of DO-254 is often so strong that equipment suppliers will go to great lengths to classify their hardware as simple even when it really is not. The fact that DO-254 does not provide a quantifiable definition of the "comprehensive

TABLE 1.3
Summary of Contents of DO-254

Section	Title	Topics
1	Introduction	Purpose/Scope, relationship to other documents, related documents, how to use DO-254, complexity, alternative methods, overview
2	System Aspects of Hardware Design Assurance	Information flow, system safety assessment processes, hardware safety assessment
3	Hardware Design Life Cycle	Life cycle processes, transition criteria
4	Planning Process	Planning process objectives and activities
5	Hardware Design Processes	Design process phases, objectives, and activities
6	Validation and Verification Process	Validation process, objectives, and activities Verification process, objectives, and activities Validation and verification methods
7	Configuration Management Process	Configuration management objectives and activities Data control categories
8	Process Assurance	Process assurance objectives and activities
9	Certification Liaison Process	Means of compliance and planning Substantiating compliance
10	Hardware Design Life Cycle Data	Plans, standards, design data, validation/verification data, acceptance test criteria, problem reports, configuration management (CM) records, process assurance records, Hardware Accomplishment Summary (HAS)
11	Additional Considerations	Previously developed hardware, COTS components, product service experience, tool assessment and qualification
Appendix A	Modulation of Hardware Life Cycle Data Based on Hardware Design Assurance Level	Life cycle data, independence
Appendix B	Design Assurance Considerations for Level A and B Functions	Functional failure path analysis (FFPA), architectural mitigation, product service experience, advanced verification methods
Appendix C	Glossary of Terms	Definitions of relevant terminology
Appendix D	Acronyms	Definitions of acronyms

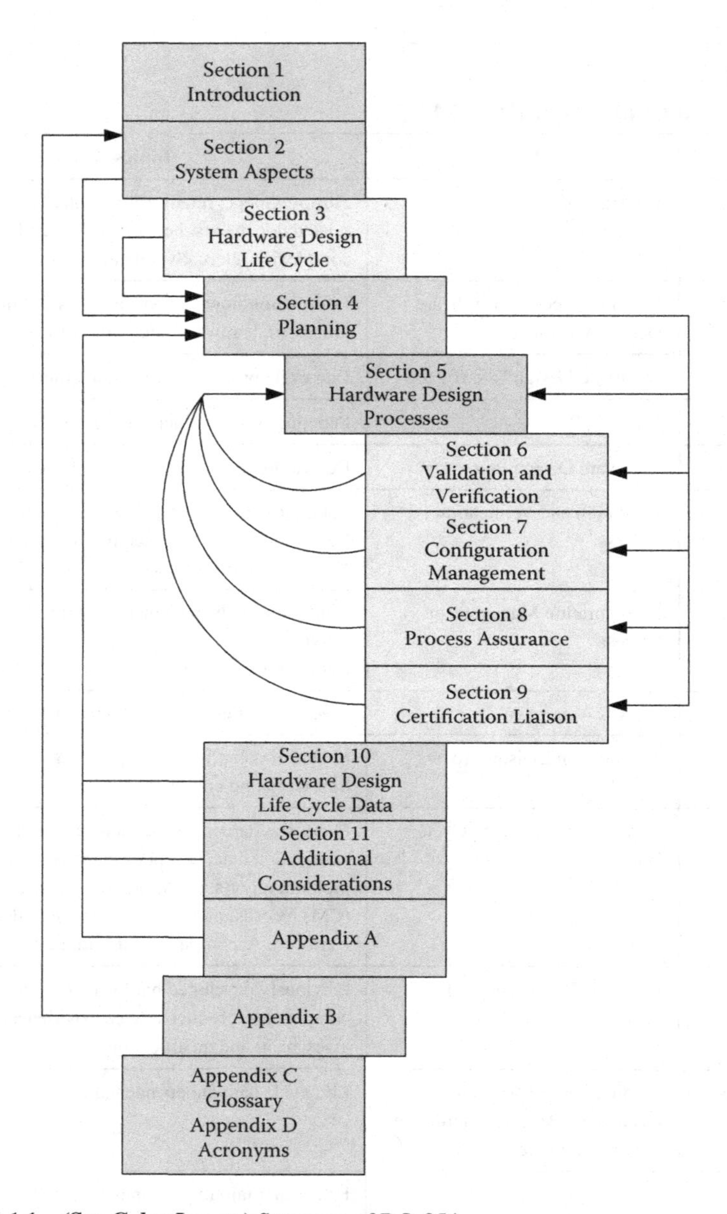

FIGURE 1.1 **(See Color Insert.)** Structure of DO-254

combination" of tests and analyses that must be performed on simple hardware just exacerbates the situation and provides the necessary ambiguity to inspire these efforts to thwart DO-254. However, FAA Order 8110.105[7] addresses the complexity issue and provides specific guidance on what constitutes the necessary comprehensive combination of tests for simple hardware, the result being that for anything other than truly simple hardware, the simple hardware approach will be either impractical or more expensive than treating the hardware as complex and complying with all of DO-254. In fact, 8110.105's verification guidance for simple hardware can be used in reverse to provide guidance on the definition of a simple device: if it is realistic or even possible to comprehensively test a hardware item according to the guidance in 8110.105, then the hardware can be correctly classified as simple.

DO-254 Section 2 describes how the processes, activities, hardware, and other data from DO-254 interface to the overall certification process. Information flow between hardware and system, and between hardware and software, is briefly discussed. Section 2 also discusses the system safety process and identifies the design assurance levels and their characteristics, as well as hardware safety assessment considerations.

DO-254 Section 3 contains a very brief discussion of the hardware design life cycle and the transition criteria that govern the movement of the life cycle from one phase or process to the next. This section identifies the life cycle processes as being the planning process, the design process, and the supporting processes. The supporting processes are further identified as being validation, verification, configuration management, process assurance, and certification liaison.

DO-254 Section 4 describes the objectives and activities of the planning process, which is the first of the design life cycle processes. Six hardware management plans and four standards are defined for hardware development: the Plan for Hardware Aspects of Certification (PHAC), Hardware Design Plan (HDP), Hardware Validation Plan (HVP), Hardware Verification Plan (HVP), Hardware Configuration Management Plan (HCMP), Hardware Process Assurance Plan (HPAP), Requirements Standards, Hardware Design Standards, Validation and Verification Standards, and Hardware Archive Standards. **Table 1.4** lists these plans and standards, where in DO-254 their contents are described, and which of the DO-254 objectives are satisfied by them.

DO-254 Section 5 describes the objectives and activities of the hardware design process. This section describes a generic design process consisting of five phases: Requirements Capture, Conceptual Design, Detailed Design, Implementation, and Production Transition. An organization's design process does not have to be the same as this process, but it should be able to satisfy all of the Section 5 objectives.

DO-254 Sections 6 through 9 describe each of the supporting processes. The objectives and activities of the validation and verification processes are described in Section 6, configuration management in Section 7, process assurance in Section 8, and certification liaison in Section 9.

DO-254 Section 6 covers the objectives and activities of the validation and verification supporting processes. Validation is the process of ensuring that derived requirements are correct and complete, and verification is the process of ensuring that the final product meets its requirements.

DO-254 Section 7 addresses the objectives and activities of the configuration management supporting process. It includes configuration items, configuration identification, baselines, problem reporting, change control, and archiving.

DO-254 Section 8 is about the process assurance supporting process, and covers its objectives and activities such as process assurance reviews and audits.

TABLE 1.4
Hardware Management Plans

Plan	Content Described in DO-254 Section	Objective Satisfied
Plan for Hardware Aspects of Certification (PHAC)	10.1.1	4.1-1, 2, 3, 4
Hardware Design Plan	10.1.2	4.1-1, 2, 3, 4
Hardware Validation Plan	10.1.3	4.1-1, 2, 3, 4 6.1.1-1
Hardware Verification Plan	10.1.4	4.1-1, 2, 3, 4 6.2.1-1
Hardware Configuration Management Plan	10.1.5	4.1-1, 2, 3, 4 7.1-3
Hardware Process Assurance Plan	10.1.6	4.1-1, 2, 4 8.1-1, 2, 3
Requirements Standards	10.2.1	4.1-2
Hardware Design Standards	10.2.2	4.1-2
Validation and Verification Standards	10.2.3	4.1-2
Hardware Archive Standards	10.2.4	4.1-2; 5.5.1-1; 7.1-1, 2

DO-254 Section 9 discusses the certification liaison supporting process, including means of compliance, planning, and how to substantiate compliance.

DO-254 Section 10 describes the content of each type of hardware life cycle data, especially the hardware management plans (PHAC, HDP, HVP, HCMP, HPAP), standards, hardware design data, verification data, and the Hardware Accomplishment Summary.

DO-254 Section 11 includes the "Additional Considerations," or in other words everything that is left over. Its contents include previously developed hardware (PDH), COTS components, product service experience, and tool assessment and qualification.

DO-254 Appendix A includes a table of all hardware life cycle data items and correlates them to their hardware control category, which in turn defines which configuration management processes and methods must be used to manage them. It also includes a definition of independence.

DO-254 Appendix B contains various verification-related techniques that must be applied with level A and B hardware. The techniques include functional failure path analysis, architectural mitigation, product service experience, and advanced verification techniques (elemental analysis, safety-specific analysis, and formal methods).

WHAT DOES IT MEAN TO ME?

Among organizations that are getting into the business of DO-254 programs, two common questions are, "What do I have to do to comply with DO-254?" and "How much will DO-254 cost?"

Compliance is obviously the simpler question to answer, and the answer is to simply adopt (and adapt to) the industry best practices in DO-254. These are summarized as follows:

- Establish a structured design process that meets the objectives defined in DO-254
  - Well-defined phases with realistic entry and exit criteria
  - Built-in peer reviews
  - Do the front-end work (requirements, traceability, plans, analysis, research, etc.) conscientiously and at the proper time
  - Honor and observe the design process (do not ignore or bypass it, even during a crisis)
- Learn how to write well-formed requirements
  - Emphasize the use of functional requirements rather than design description (implementation) requirements
  - Write them early
  - Understand what a derived requirement is
  - Learn how to document validation and traceability data as part of requirements capture
- Establish high integrity validation and verification processes and methods, including:
  - An independent peer review process for levels A and B
  - Peer reviews of data and documents
  - Requirement reviews to validate derived requirements
  - Learn how to conduct requirements-based verification
  - Learn how to write effective and optimized test cases
  - Learn how to conduct robustness verification
  - Simulations—functional and post-layout timing (or static timing analysis)
  - Elemental analysis—code coverage
  - Hardware test—in the actual flight hardware
  - Acquire tools to support verification
- Establish a configuration management infrastructure, including:
  - Problem reporting
  - Document/data control
  - Document/data release
  - Backups/archives
  - Refresh of backup media
  - Retention of data for life of use of the equipment
  - Retention of tools and test equipment for life of use of the equipment

- Establish a component management process—be prepared for parts obsolescence
- Establish a process assurance role or department
  - Perform audits, review
  - Track deviations
  - Audit transition criteria
- Be prepared for audits of off-shore or subcontracted work
  - May need a sub-supplier management plan
  - May need on-site stage of involvement (SOI) audits
  - Close coordination with configuration management and control of data
  - Close coordination with process assurance
  - Technical oversight
- Interface with customer and/or airframer
  - Establish and interpret requirements
- Interface with certification authority
  - FAA and/or designee
  - SOI audits
- Organize vast amounts of data
- Write lots of documentation and reports
- Cultural changes
  - Use the CM system
  - Use change control
  - Comply with processes—no more informal engineering
  - Transparency
  - Accountability
- Learn new rules for component selection
  - Settle for less than the leading edge technology
  - Embrace more tried and true technology, tools, methods
  - New is ok, but may require more verification or justification

Obviously the list seems long, but for most companies, compliance might require adaptation (or not) but should never be impossible. Changing the engineering culture—if it is needed, since some companies have already discovered the benefits of using the best practices in DO-254—is often the hardest part of complying with DO-254, but with time it can be done.

The second question is somewhat dependent on the answer to the first one and is considerably harder to answer, mainly because there simply is not a good quantifiable answer to it. Some people try to attach a percentage escalation factor to a project, for example adding on 30 percent to a project's non-DO-254 cost to account for DO-254 compliance, but in reality the cost is going to be determined by a company's DO-254 "gap" (how much its existing processes fall short of DO-254) and how agreeable the company and its personnel are to changing their ways and learning new tricks, i.e., how willing they are to close that gap. A company whose processes are reasonably consistent with DO-254 can conceivably incur significant additional costs simply because its personnel resist DO-254 rather than embrace it. Likewise, a company that has little or none of the DO-254 infrastructure can conceivably comply with DO-254 with minimal inconvenience and cost if the company and its people are open to learning and using the processes in it. The amount of discomfort and cost associated with DO-254 is almost entirely self-determined, and the lesson that "the most expensive path to DO-254 compliance is the one that tries to go around it" can be a difficult and expensive one to learn.

For convenience, references to DO-178B and DO-178C in this book will simply use DO-178 when the discussion applies equally to both versions of DO-178. The reference to DO-178B will be used when reference to the earlier version is intended. Similarly, references to SAE ARP4754 and SAE ARP4754A will use ARP4754 when the discussion applies equally to both versions of ARP4754.

REFERENCES

1. RTCA DO-254, *Design Assurance Guidance for Airborne Electronic Hardware*, RTCA Inc., Washington, D.C., 2000. This document is also known as EUROCAE ED-80.
2. Advisory Circular Number 20-152, RTCA, INC., DOCUMENT RTCA/DO-254, DESIGN ASSURANCE GUIDANCE FOR AIRBORNE ELECTRONIC HARDWARE, Federal Aviation Administration, June 2005.
3. SAE ARP4754, *Certification Considerations for Highly-Integrated or Complex Aircraft Systems*, Warrendale, PA: SAE, 1996.
4. SAE ARP4754A, *Guidelines for Development of Civil Aircraft and Systems*, Warrendale, PA: SAE, 2010.
5. RTCA DO-178B, *Software Considerations in Airborne Systems and Equipment Certification*, RTCA Inc., Washington, D.C., 1992. This document is also known as EUROCAE ED-12B.
6. RTCA DO-178C, *Software Considerations in Airborne Systems and Equipment Certification*, RTCA Inc., Washington, D.C., 2011. This document is also known as EUROCAE ED-12C.
7. Order 8110.105 CHG 1, *Simple and Complex Electronic Hardware Approval Guidance*, Federal Aviation Administration, dated September 23, 2009.

FURTHER INFORMATION

1. The Federal Aviation Administration Web page: www.faa.gov
2. The RTCA Web page: www.rtca.org
3. The SAE Web page: www.sae.org
4. EASA Web page: www.easa.europa.eu
5. FAA Regulatory and Guidance Library (RGL) Web page: rgl.faa.gov
6. CAST Position Papers Web page: www.faa.gov/aircraft/air_cert/design_approvals/air_software/cast/cast_papers/

2 Regulatory Background

In the United States, the Code of Federal Regulations (CFR)[1] is the codification of the rules published in the Federal Register by the departments and agencies of the federal government. The Code of Federal Regulations is divided into 50 titles that represent areas subject to federal regulation. The 50 subject matter titles contain one or more volumes that are updated annually. Each title is divided into chapters, and each chapter typically bears the name of the issuing agency. Each chapter is further subdivided into parts that cover specific regulatory areas. Large parts may be subdivided into subparts; parts are then organized in sections.

Title 14 of the CFR covers regulations for aeronautics and space. Within Title 14 there are six Chapters spread over five Volumes. The Volumes, Chapters, and Parts are organized as shown in **Table 2.1**.

Chapter one of Title 14 has three subchapters that define the contents of Part 1–59. The subchapters are as follows:

- Subchapter A—Definitions
  - Part 1 Definitions and Abbreviations
  - Part 3 General Requirements
- Subchapter B—Procedural Rules
  - Part 11 General Rulemaking Procedures
  - Part 13 Investigative and Enforcement Procedures
  - Part 14 Rules Implementing the Equal Access to Justice Act of 1980
  - Part 15 Administrative Claims Under Federal Tort Claims Act
  - Part 16 Rules of Practice for Federally Assisted Airport Enforcement Proceedings
  - Part 17 Procedures for Protests and Contract Disputes
- Subchapter C—Aircraft
  - Part 21 Certification Procedures for Products and Parts
  - Part 23 Airworthiness Standards: Normal, Utility, Acrobatic, and Commuter Category Airplanes
  - Part 25 Airworthiness Standards: Transport Category Airplanes
  - Part 26 Continued Airworthiness and Safety Improvements for Transport Category Airplanes
  - Part 27 Airworthiness Standards: Normal Category Rotorcraft
  - Part 29 Airworthiness Standards: Transport Category Rotorcraft
  - Part 31 Airworthiness Standards: Manned Free Balloons
  - Part 33 Airworthiness Standards: Aircraft Engines

TABLE 2.1
Title 14 Code of Federal Regulations

Title	Volume	Chapter	Part	Regulatory Entity
Title 14 Aeronautics and Space	1	I	1–59	Federal Aviation Administration, Department of Transportation
	2		60–109	
	3		110–199	
	4	II	200–399	Office of the Secretary, Department of Transportation (Aviation Proceedings)
		III	400–1199	Commercial Space Transportation, Federal Aviation Administration, Department of Transportation
	5	V	1200–1299	National Aeronautics and Space Administration
		VI	1300–1399	Air Transportation System Stabilization

- Part 34 Airworthiness Standards: Fuel Venting and Exhaust Emission Requirements for Turbine Engine Powered Airplanes
- Part 35 Airworthiness Standards: Propellers
- Part 36 Noise Standards: Aircraft Type and Airworthiness Certification
- Part 39 Airworthiness Directives
- Part 43 Maintenance, Preventive Maintenance, Rebuilding and Alteration
- Part 45 Identification and Registration Marking
- Part 47 Aircraft Registration
- Part 49 Recording of Registration Titles and Security Documents
- Part 50-59 [Reserved]

Figure 2.1 shows the structure of the Code of Federal Regulations, Title 14 for Aeronautics and Space, Part 21/23/25/27/29 for parts, aircraft, and engines, regulation 1301 for function and installation, and regulation 1309 for equipment, systems, and installation.

A particular regulation is referenced by Title, Part, Subchapter, and Subpart. The regulations pertaining to systems and equipment on Part 25 transport aircraft are referenced:

- 14 CFR 25.1301[2]
- 14 CFR 25.1309[3]

Federal Aviation Regulations have an amendment level to identify the date and identification of the most recent change to the regulation. In general, aircraft certification programs for new aircraft or changes to existing aircraft use the most recent amendment level of the regulations applicable to the program. In some instances

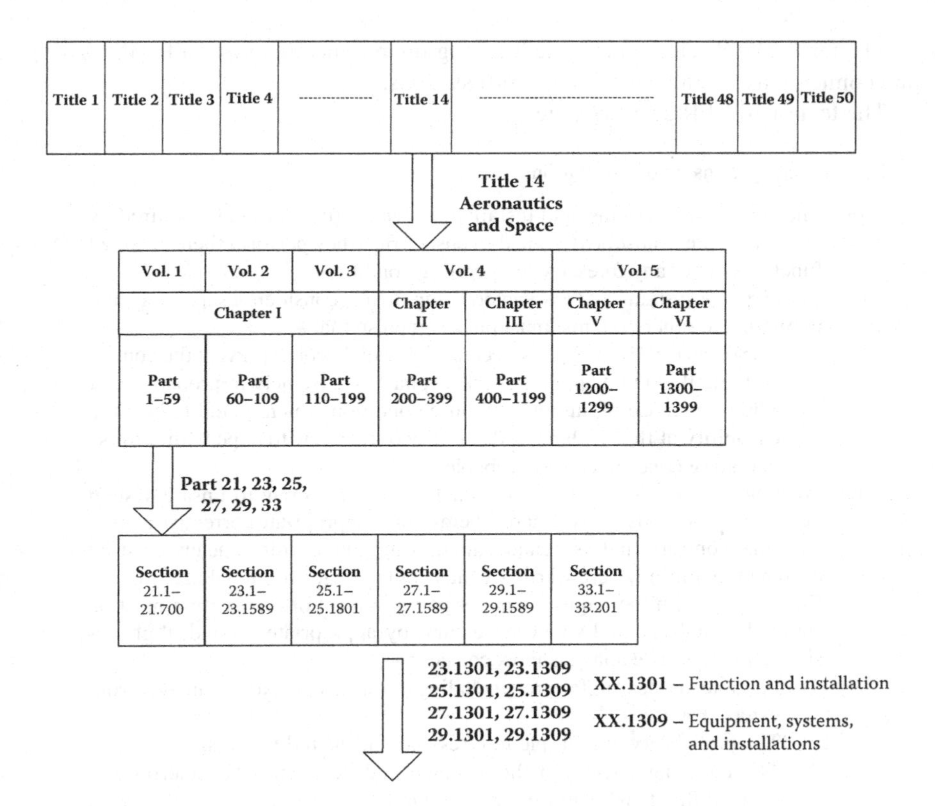

FIGURE 2.1 CFR Structure

programs such as aircraft derivative models or changes to an existing model aircraft can use the regulations and amendment level from original approval or certification basis.

The following paragraphs show the basic text of FARs 25.1301 and 25.1309. While other versions and amendments of this FAR exist, the point here is to show the basic intent of the text. Foreign certification authorities such as the European Aviation Safety Agency (EASA) use a set of Certification Specification (CS) that follow the numbering, and largely, the intent or same text, as the equivalent FAR.

The text of 14 CFR 25.1301 states:

Function and installation.

(a) Each item of installed equipment must—
  (1) Be of a kind and design appropriate to its intended function;
  (2) Be labeled as to its identification, function, or operating limitations, or any applicable combination of these factors;
  (3) Be installed according to limitations specified for that equipment; and
  (4) Function properly when installed.

[(b) EWIS must meet the requirements of subpart H of this part.]

Amendment 25-123, Effective 12/10/07

The term EWIS refers to electrical wiring interconnection system for the wiring and connections to, and between, aircraft systems.

The text of 14 CFR 25.1309 states:

Equipment, systems, and installations.

(a) The equipment, systems, and installations whose functioning is required by this subchapter, must be designed to ensure that they perform their intended functions under any foreseeable operating condition.
(b) The airplane systems and associated components, considered separately and in relation to other systems, must be designed so that—
 (1) The occurrence of any failure condition which would prevent the continued safe flight and landing of the airplane is extremely improbable, and
 (2) The occurrence of any other failure condition which would reduce the capability of the airplane or the ability of the crew to cope with adverse operating conditions is improbable.
(c) Warning information must be provided to alert the crew to unsafe system operating conditions, and to enable them to take appropriate corrective action. Systems, controls, and associated monitoring and warning means must be designed to minimize crew errors which could create additional hazards.
(d) Compliance with the requirements of paragraph (b) of this section must be shown by analysis, and where necessary, by appropriate ground, flight, or simulator tests. The analysis must consider—
 (1) Possible modes of failure, including malfunctions and damage from external sources.
 (2) The probability of multiple failures and undetected failures.
 (3) The resulting effects on the airplane and occupants, considering the stage of flight and operating conditions, and
 (4) The crew warning cues, corrective action required, and the capability of detecting faults.
[(e) In showing compliance with paragraphs (a) and (b) of this section with regard to the electrical system and equipment design and installation, critical environmental conditions must be considered. For electrical generation, distribution, and utilization equipment required by or used in complying with this chapter, except equipment covered by Technical Standard Orders containing environmental test procedures, the ability to provide continuous, safe service under foreseeable environmental conditions may be shown by environmental tests, design analysis, or reference to previous comparable service experience on other aircraft.]
[(f) EWIS must be assessed in accordance with the requirements of Section 25.1709.]

Amendment 25-123, Effective 12/10/07

The certification basis for a project or an aircraft program is the applicable airworthiness requirements as established in 14 CFR 21.17[4] (original certification) and 14 CFR 21.101[5] (change to a type certificate), as appropriate; special conditions; equivalent level of safety findings; requirements per 14 CFR 21.21(b)(2)[6]; and exemptions applicable to the product to be certificated. The certification basis specifies the applicable regulations and their respective amendment level at the time of application for a certificate. An application for type certification of a transport category aircraft (Part 25) is effective for five years. An application for any other type certificate is effective for three years, unless an applicant shows at the time of application that their product requires a longer period of time for design, development, and testing, and the FAA approves a longer period. The applicant is the party that applies to the FAA for a type certification, supplemental type certificate, or an amended type certificate. A type certificate (TC) is the certification for a new type of aircraft. A supplemental type certificate (STC) is the certification for modification to existing aircraft. An amended type certificate (ATC) is FAA approval to modify an aircraft design from its original design. An amended type certificate approves not only the modification, but also how that modification affects the original design.

MEANS OF COMPLIANCE

The FAA publishes additional information that explains an acceptable method to comply with the certification basis and thus to the applicable regulations (FAR). These publications are known as Advisory Circulars (AC). The Advisory Circulars contain information known as a **means of compliance** (MoC) to a FAR. Advisory Circulars are identified by the regulation they are associated with. For example, AC 25.1309[7] explains a means of compliance to regulation 14 CFR 25.1309. Advisory Circulars may also contain explanations of regulations and other guidance materials, best practices, or information useful to the aviation industry. Advisory Circulars can also provide guidance, methods, procedures, and practices for complying with Federal Aviation Regulations and requirements.

The most commonly used means of compliance to a FAR are:

- Engineering evaluation
  - Compliance statement
  - Design review
  - Calculation/analysis
  - Safety assessment
- Tests
  - Laboratory test
  - Ground test on aircraft
  - Flight test
  - Simulation
- Inspection
  - Conformity inspection
  - Design inspection
- Equipment qualification

Engineering evaluation for demonstration of compliance to a regulation includes compliance statement, design review, calculation/analysis, and safety assessment. A compliance statement is a formal statement that a design complies with a regulation and may include compliance by similarity. A design review is used when compliance is based on a review of data, descriptions, or drawings. Calculation or analysis is used when compliance is demonstrated by an engineering analysis, calculation, or report. Safety assessment is used when compliance is demonstrated by a safety analysis such as probability analysis.

Tests for demonstration of compliance to a regulation include laboratory tests, ground tests with an aircraft, flight tests with an aircraft, and simulation of aircraft functions. Laboratory test is used when compliance is demonstrated by testing in a lab. Ground test is used when compliance is demonstrated by aircraft testing conducted on the ground. Flight test is used when compliance is demonstrated with aircraft flight testing. Simulation is used when compliance is demonstrated by flight or computer model simulation or with the use of a representative mockup of the equipment.

Inspection for demonstration of compliance to a regulation includes conformity inspection and design inspection. Conformity inspection is used when compliance is demonstrated by a review of the equipment against its approved and released drawings. Design inspection is used when compliance is demonstrated by an inspection of the design as it is installed on the aircraft.

Equipment qualification is used when compliance is demonstrated by environmental qualification testing, such as RTCA/DO-160.[8]

The means of compliance allow the applicant to "show compliance" to a regulation. Once the compliance has been demonstrated, the regulatory authority or designee may then make a finding of compliance to the regulation.

Many of the methods for demonstrating compliance to the regulations are not well suited or practical for airborne electronic hardware, especially circuit cards or devices such as field programmable gate arrays (FPGAs), application specific integrated circuits (ASICs), or complex programmable logic devices (CPLDs). DO-254 was written to provide applicants and developers with suitable methods and techniques to show compliance to the FARs.

Figure 2.2 shows the Advisory Circulars for FARs pertaining to systems, equipment, and electronic parts.

AC 25.1309 includes several objectives based on fail-safe system design principles. When considering a system and its operation, failure objectives need to include:

- The failure of any single element, component, or connection during any one flight.
- Any single failure of an element, component, or connection should not prevent safe flight and landing.
- Catastrophic failures must be extremely improbable.
- Common cause analysis should be used to ensure that single failures do not adversely affect more than one channel in a redundant system (i.e., with multiple channels). These single failures do not adversely affect more than one system performing equivalent aircraft functions.

AC 25.1309 also includes techniques to assess failures and their impacts. These assessments are supported with qualitative and quantitative analysis. A functional hazard assessment (FHA) is a qualitative assessment to identify and classify failures. The FHA is used in early design stages to assess various system architectures and design for suitability. The FHA, as will be described later, is also used

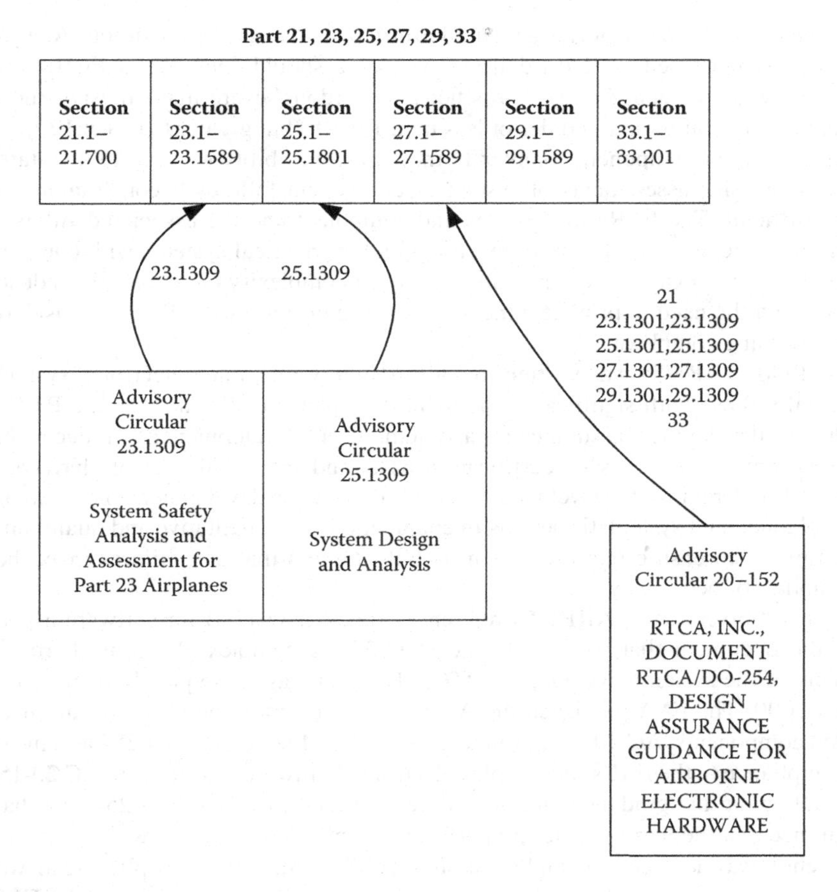

FIGURE 2.2 Advisory Circulars

to determine the DAL associated with a function and its electronic hardware. In Aerospace Recommended Practice (ARP) 4754A, the aircraft level function design assurance level is the FDAL, the software and/or airborne electronic hardware has the resultant item design assurance level (IDAL). A reliability or fault tree analysis is a quantitative assessment of failures to show that their probability is commensurate, i.e., inversely proportional, to their severity.

In accordance with AC 25.1309, catastrophic hazards must be shown to be extremely improbable. Analysis of catastrophic hazards should include qualitative and quantitative assessment. Failures classified as major hazards must be shown to be improbable. Analysis of major hazards includes qualitative and sometimes a quantitative assessment.

The AC goes on to define extremely improbable failures as those having a probability on the order of 1×10^{-9} or less, improbable failures as those having a probability on the order of less than 1×10^{-5} but greater than 1×10^{-9}, and probable failures as those having a probability on the order of greater than 1×10^{-5}.

Updates to these categories added a hazardous/severe-major condition. Analysis of failures classified as hazardous/severe-major should include qualitative and quantitative assessment. Failures classified as hazardous/severe-major must be shown to have a probability on the order of less than 1×10^{-5} but greater than 1×10^{-7}.

As system or component complexity increases, the ability to perform qualitative and quantitative assessments of system or component failures becomes more and more difficult. The FARs for systems and equipment and the associated Advisory Circular came about in the era of electrical and mechanical systems with comparatively low complexity. As system and component complexity increased, the industry recognized the need to provide a means of compliance to the FARs that considered the increasing complexity.

In 1996, ARP4754 was written to address highly integrated electronic systems, especially those with significant functionality implemented in software. ARP4754 addresses development assurance for a system, where development assurance results in a system that satisfies its certification basis and from which errors have been detected and removed. Development assurance uses a development methodology with planned and systematic actions in conjunction with qualitative and quantitative assessments to ensure that the system is fail-safe and that any failures meet their probability of occurrence.

The 1996 version of ARP4754 references RTCA/DO-178B for software aspects and the document that would become DO-254 for complex electronic hardware aspects of the system development. RTCA/DO-254 was subsequently published in April 2000. The FAA published the Advisory Circular for complex custom microcoded components, AC 20-152, in June 2005. AC 20-152 defines DO-254 as a means of compliance to the FARs for complex electronic hardware. As stated in AC 20-152, "By following the guidance and procedures outlined in RTCA/DO-254, you have assurance that the hardware design performs its intended functions within the environment it was designed for, and the assurance of meeting all the applicable airworthiness requirements."[9] Notice that this language wraps back around to the FARs and the Advisory Circulars:

- FAR 25.1301 states that equipment must operate properly and FAR 25.1309 states that equipment must perform its intended function.
  - AC 20-152 states that if DO-254 is used, then there is assurance that complex electronic hardware will perform its intended function.
- FAR 25.1309 states that equipment must perform its intended function under all foreseeable operating conditions.
  - AC 20-152 states that if DO-254 is used, then there is assurance that complex electronic hardware will perform its intended function in the environment for which it was designed.
- AC 25.1309 states that qualitative and quantitative analysis should be used for catastrophic and hazardous/severe-major failure conditions. The AC also gives probabilities for all categories of failure conditions. AC 25.1309 is a means of compliance to FAR 25.1309.
  - AC 20-152 states that if DO-254 is used then there is assurance of meeting applicable airworthiness requirements.

Advisory Circular 20-152, published in June 2005, addresses RTCA/DO-254, Design Assurance Guidance for Airborne Electronic Hardware for 14 CFR Parts 21, 23, 25, 27, 29, and 33. DO-254 can be used for compliance to Federal Aviation Regulations for:

- Products and Parts
- Normal, Utility, Acrobatic, and Commuter Category Airplanes
- Transport Category Airplanes
- Normal Category Rotorcraft
- Transport Category Rotorcraft
- Aircraft Engines

GUIDANCE MATERIALS

Guidance is information that explains how to comply with certification requirements and aviation regulations. FAA Advisory Circular 20-152 recognizes the guidance in DO-254 for complex custom micro-coded components with hardware design assurance levels of A, B, and C (i.e., the IDAL). The types of components covered by AC 20-152 include ASICs, PLDs, FPGAs, and similar components.

The guidance in DO-254 is applicable to line replaceable units (LRU), circuit cards or circuit boards, programmable logic devices including CPLDs, ASICs, and FPGAs, and COTS components. While DO-254 has a broad scope, AC 20-152 limits the scope to complex custom micro-coded components. The guidance in DO-254 represents industry consensus on design assurance for airborne electronic hardware. DO-254 also incorporates best practices for design assurance identified by the aviation and electronics industry.

Guidance in DO-254 covers a variety of topics including:

- Hardware standards
- Hardware design life cycle data
- Additional design assurance techniques for design assurance (IDAL) A and B functions
- Previously developed hardware
- Tool assessment and qualification
- Use of COTS components
- Product service experience
- Hardware safety assessment
- Design assurance strategy, including consideration for design assurance (IDAL) A and B functions
- Planning process
- Requirements capture
- Conceptual design
- Detailed design
- Implementation
- Production transition
- Validation process

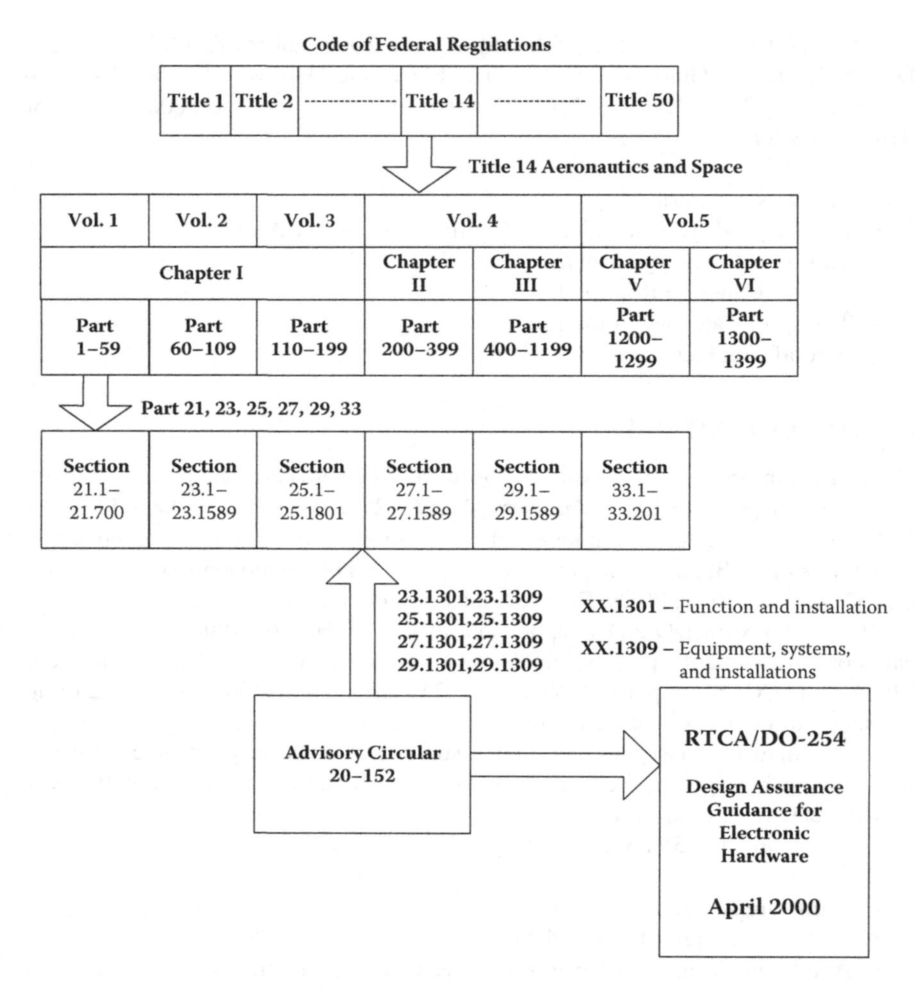

FIGURE 2.3 DO-254 Relationship to CFR

- Verification process including tests and reviews
- Configuration management
- Process assurance
- Certification liaison and proposed means of compliance

The path to DO-254 from the Code of Federal Regulations is shown in **Figure 2.3**.

ISSUE PAPERS

The Federal Aviation Administration (FAA) uses Issue Papers to provide a structured means of accomplishing the necessary steps in the type certification and type validation processes. Type certification includes projects for type certificates, amended type certificates, type design changes, supplemental type certificates, and amended supplemental type certificates. Issue Papers provide a means for describing and tracking the resolution of significant technical, regulatory, and administrative issues that occur during a project. The Issue Paper process establishes a formal communication for significant issues between the applicant, a foreign civil aviation authority if applicable, and the FAA.

For type certification projects, Issue Papers are useful tools for keeping an unbiased uniform certification approach between applicants. Issue Papers also form a valuable reference for future type certification programs and for development of regulatory changes. By describing significant or precedent setting technical decisions and the rationales employed, Issue Papers can become reference material.

The Issue Paper process is documented in FAA Order 8110.112[10] titled Standardized Procedures for Usage of Issue Papers and Development of Equivalent Levels of Safety Memorandums. The FAA maintains a Transport Airplane Issues List for issues applicable to international validation and domestic certification projects on the FAA Web site.

The list identifies the subject "Assurance of Simple and Complex Electronic Hardware" for inclusion in certification programs. The description states that a means of compliance Issue Paper may be needed for most aircraft programs and modification projects for airborne systems containing electronic hardware components.

Current FAA certification programs use an Issue Paper to cover topics relevant to airborne electronic hardware. Contents of these Issue Papers are proprietary to the applicant. Foreign certification authorities have similar processes; e.g., EASA has a similar process that uses certification review items (CRI).

ORDERS

The FAA uses Orders to provide information to the managers and staff of the FAA Aircraft Certification Service, including designated engineering representatives (DER), and organizations associated with the certification process described in Title 14 of the Code of Federal Regulations. Order 8110.105 applies to PLDs and is titled *Simple and Complex Electronic Hardware Approval Guidance.*

Order 8110.105 supplements RTCA/DO-254 and gives guidance for approving both simple and complex custom micro-coded components. Topics covered, and the associated paragraphs in 8110.105, are:

- Reviews and FAA involvement
  - Chapter 2: How to review simple electronic hardware (SEH) and complex electronic hardware (CEH)
  - Chapter 3: How much FAA involvement should apply to hardware projects
- Topics for SEH and CEH
  - Chapter 4–2: Modifiable components
  - Chapter 4–3: Certification plan
  - Chapter 4–4: Validation processes
  - Chapter 4–5: Configuration management
  - Chapter 4–6: Assessing and qualifying tools

- Chapter 4–7: Approving hardware changes in legacy systems using RTCA/DO-254
- Chapter 4–8: Acknowledging compliance to RTCA/DO-254 for TSO approvals that don't reference RTCA/DO-254
- Chapter 4–9: COTS intellectual property
- Topics for SEH
  - Chapter 5–2: Verification processes
  - Chapter 5–3: Traceability
- Topics for CEH
  - Chapter 6–2: Verification processes
  - Chapter 6–3: Traceability

Chapter 2 of Order 8110.105 describes reviews as part of the certification liaison process. Reviews may be conducted on-site, at the applicant or supplier's facility, or may be conducted as a desk review. The four reviews, known as SOI reviews, are as follows:

- SOI #1—Hardware planning review
- SOI #2—Hardware design review
- SOI #3—Hardware validation and verification review
- SOI #4—Final review

SOI #1 is conducted when most of the plans and standards are complete and reviewed. SOI #2 is typically conducted when at least 50 percent of the hardware design data (requirements, design, and implementation) is complete and has been reviewed. SOI #3 is typically conducted when at least 50 percent of the hardware validation and verification data is complete and has been reviewed. SOI #4 occurs after the final hardware build and verification are complete, a hardware conformity review is done, and the application(s) is ready for formal system approval.

The depth and extent of FAA involvement in a project, including determination of when and where to conduct the SOI reviews, is determined by the hardware design assurance level and a score based on relevant criteria. Level A and B hardware require high or medium FAA involvement, Level C hardware requires medium or low FAA involvement, and Level D hardware requires low FAA involvement. The score derived from relevant criteria can range from 0 to 207. A score of less than 80 results in high FAA involvement for Level A and B, medium FAA involvement for Level C, and low FAA involvement for Level D hardware. A score of between 80 and 130 results in high FAA involvement for Level A, medium FAA involvement for Level B and C, and low FAA involvement for Level D hardware. A score of greater than 130 results in medium FAA involvement for Level A and B, and low FAA involvement for Level C and D hardware. The FAA involvement could be direct participation by an FAA engineer from an Aircraft Certification Office (ACO) or by a DER assigned and delegated for the project. Compliance organizations can use Authorized Representatives (AR) or Unit Members (UM) as the designated authority to perform oversight and reviews.

Stage of Involvement audits use the content of the Airborne Electronic Hardware Review Job Aid[11] as a reference tool during the reviews. While the Job Aid is not intended to be used as a checklist, it often does serve as the official checklist during a review. Either way, it would behoove applicants and developers to be familiar with the Job Aid questions. The Job Aid questions can form the basis for project review forms or can be used to conduct internal dry run reviews before a formal FAA review is held.

The Job Aid also includes a description of tasks to be performed before, during, and after a hardware review in Part 2; a list of activities and questions to be considered during a review in Part 3; and an approach to the Findings and Observations to DO-254 objectives in Part 4.

REFERENCES

1. Code of Federal Regulations, United States Government Printing Office, electronic version available at: http://www.ecfr.gov
2. Code of Federal Regulations, Title 14: Aeronautics and Space, PART 25—AIRWORTHINESS STANDARDS: TRANSPORT CATEGORY AIRPLANES, Subpart F—Equipment, 25.1301 Function and installation.
3. Code of Federal Regulations, Title 14: Aeronautics and Space, Part 25—AIRWORTHINESS STANDARDS: TRANSPORT CATEGORY AIRPLANES, Subpart F—Equipment, 25.1309 Equipment, systems, and installations.
4. Code of Federal Regulations, Title 14: Aeronautics and Space, PART 21—CERTIFICATION PROCEDURES FOR PRODUCTS AND PARTS, Subpart B—Type Certificates, 21.17 Designation of applicable regulations.
5. Code of Federal Regulations, Title 14: Aeronautics and Space, PART 21—CERTIFICATION PROCEDURES FOR PRODUCTS AND PARTS, Subpart D—Changes to Type Certificates, 21.101 Designation of applicable regulations.
6. Code of Federal Regulations, Title 14: Aeronautics and Space, PART 21—CERTIFICATION PROCEDURES FOR PRODUCTS AND PARTS, Subpart B—Type Certificates, 21.21 Issue of type certificate: normal, utility, acrobatic, commuter, and transport category aircraft; manned free balloons; special classes of aircraft; aircraft engines; propellers.
7. Advisory Circular, AC 25.1309-1A—System Design and Analysis, Federal Aviation Administration, June 1988.
8. RTCA DO-160F, ENVIRONMENTAL CONDITIONS AND TEST PROCEDURES FOR AIRBORNE EQUIPMENT, RTCA Inc., Washington, D.C., 2000.
9. Advisory Circular Number 20-152, RTCA, INC., DOCUMENT RTCA/DO-254, DESIGN ASSURANCE GUIDANCE FOR AIRBORNE ELECTRONIC HARDWARE, Federal Aviation Administration, June 2005, p. 1.
10. Order 8110.112, *Standardized Procedures for Usage of Issue Papers and Development of Equivalent Levels of Safety Memorandums*, Federal Aviation Administration, June 2010.
11. *Conducting Airborne Electronic Hardware Reviews Job Aid*, Aircraft Certification Service, Federal Aviation Administration, February 2008.

3 Planning

DO-254 development takes place within the context of ARP4754 processes. The certification aspects of airborne electronic hardware are encapsulated by the development of the system in which the electronic hardware is used. Similarly, the system development is performed within the context of a certification program for an aircraft, a modification or update to an aircraft, or in the case of a technical standard order (TSO), the TSO approval process.

Figure 3.1 shows the airborne electronic hardware within a system and the system within the aircraft program concept.

The system and safety processes start before, or along with, the DO-254 planning processes. The system process outputs that factor into planning include:

- System description
- Hardware description

The safety process outputs that factor into planning include:

- Top-level hazards from the functional hazard assessment that are the drivers for the design assurance level.
- Preliminary system safety assessment that establishes the function design assurance level at the aircraft level and the item design assurance level for the hardware.
- Functional failure path analysis that establishes which failures the hardware participates in and the corresponding classification for the hazard associated with the failure.

Figure 3.2 shows the relationship between system and safety ARP4754 processes and electronic hardware DO-254 processes. The Formal Test phase included in Figure 3.2 is not mandated by DO-254.

The aircraft or TSO certification program will determine the certification basis for the program. The certification basis is the applicable Federal Aviation Regulations and their respective amendment level and any project specific FAA Issue Papers. The certification basis is typically defined in the system level certification plan. The certification basis for airborne electronic hardware will have the same certification basis as the equipment in which it is used. Certification programs conducted in concert with foreign certification agencies may have additional issues levied to meet the requirements of other countries. The outputs of the aircraft, system, and safety processes become inputs for the formulation of the project-specific Plan for Hardware

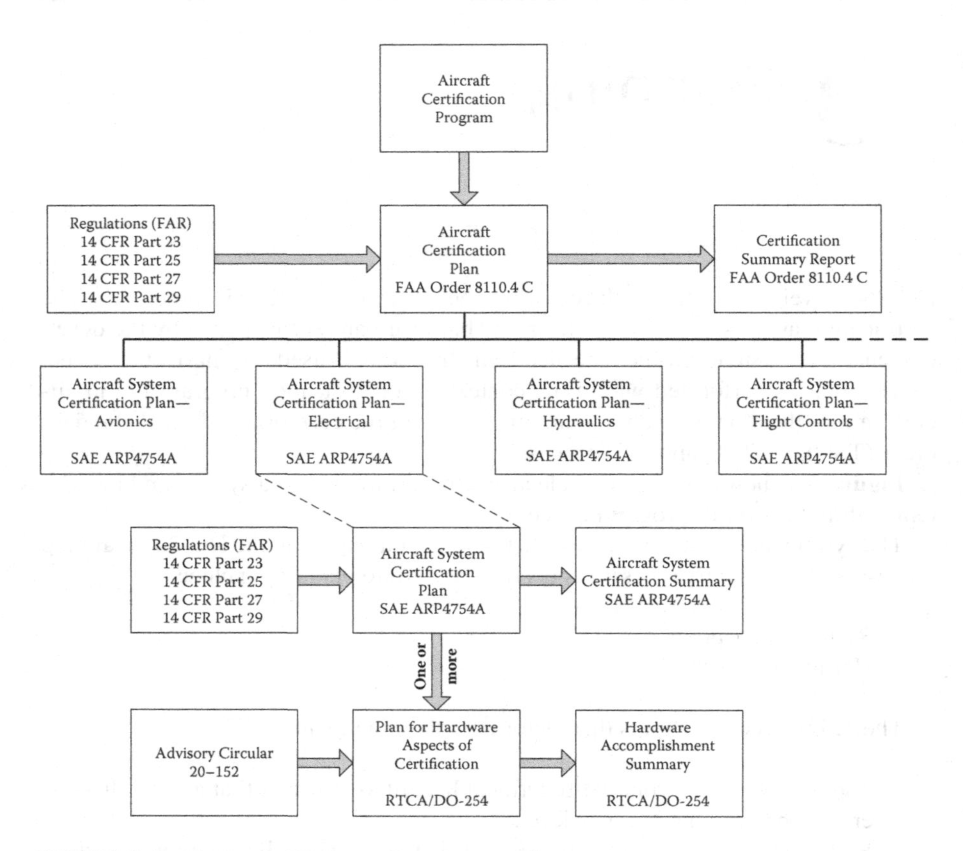

FIGURE 3.1 Electronic Hardware Development Context

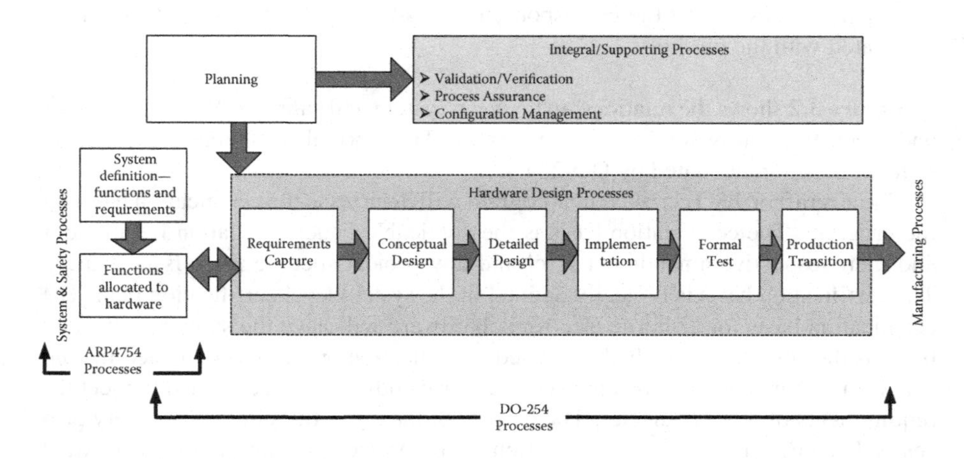

FIGURE 3.2 ARP4754 and DO-254 Processes

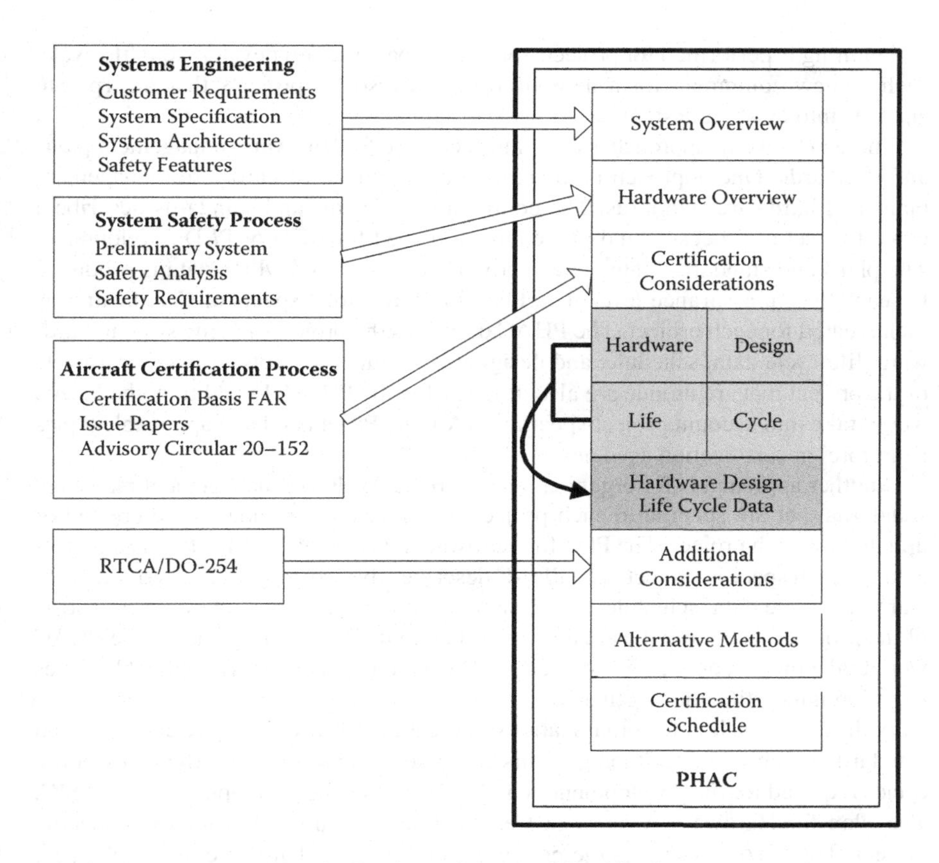

FIGURE 3.3 Inputs to Plan for Hardware Aspects of Certification

Aspects of Certification, or PHAC. The aircraft and system process outputs and PHAC inputs are depicted in **Figure 3.3**.

Project planning begins with a combination of several aspects—the idea for the design, gathering the processes and tools to accomplish the design, establishing the resources and organization to accomplish the tasks, evaluating customer-specific requirements, and evaluating project-specific aspects including the certification basis.

Planning considers hardware development from conception all the way through development and into production. Planning also includes production aspects after the design is complete to account for parts obsolescence or parts change. Planning should be a horizon-to-horizon vision of the project to ensure that all phases correctly describe the processes, methodologies, and tools used.

The planning can include a trade study or design trade-off that evaluates whether a particular design or implementation technology can be realized. Planning an FPGA design with a device that has been used in other programs and that uses the same design and verification tools will have less planning or preparation than embarking on a project that uses multiple intellectual cores developed on an FPGA, converts the design into an ASIC for production, and uses new verification tools and techniques.

Planning is performed for projects that encompass the full development life cycle, such as new equipment for a new aircraft, and also for projects that incorporate changes into existing designs.

There are several approaches companies can use for hardware management plans and standards. One approach is to create a common set of hardware management plans and hardware standards. The common set of plans and standards describe a consistent set of processes and procedures followed for AEH or PLD development. The plans and standards state which activities are used and which data is produced for each design assurance level. The Plan for Hardware Aspects of Certification is then created for each project. The PHAC describes the project-specific system, hardware, life cycle data, schedule, and design and verification tools. Any other aspects of the project that are unique are also described in the PHAC. In addition, the PHAC would take into account project-specific FAA Issue Papers and any applicable topics from foreign certification agencies.

Another approach is for organizations to create hardware management plans and standards that are specific to each project. The plans and standards are created or updated for each project. The Plan for Hardware Aspects of Certification is also created for each specific project. The PHAC describes the project-specific system, hardware, life cycle data, schedule, and design and verification tools. Any other aspects of the project that are unique are also described in the PHAC. In addition, the PHAC would take into account project-specific FAA Issue Papers and any applicable topics from foreign certification agencies.

A third approach is for organizations to create hardware management plans and standards specific to a technology. This could be the case for the differences in life cycles required for the development of an FPGA versus the development of an ASIC. The Plan for Hardware Aspects of Certification is then created for each specific project. The PHAC describes the project-specific system, hardware, life cycle data, schedule, and design and verification tools. Any other aspects of the project that are unique are also described in the PHAC. In addition, the PHAC would take into account project-specific FAA Issue Papers and any applicable topics from foreign certification agencies.

A fourth approach is for organizations to create hardware management plans and standards that are specific to development approaches. In this case the plans would focus on the unique requirements that are specific to different development approaches, such as new product development, modifications to previously developed hardware, reuse of previously developed hardware, etc. As with other approaches the PHAC would be specific to the project, and would describe the system, hardware, life cycle data, schedule, processes, and tools that are unique to each development approach, as well as the project-specific Issue Papers and topics from foreign certification agencies.

Planning also takes into account the groups, organizations, and even companies involved in the development of airborne electronic hardware. Using the tools, processes, procedures, and personnel of an experienced DO-254 team that is located within one facility of a company is quite different from planning for spreading the life cycle and activities across different groups, different companies, and groups located in different countries, or outsourcing some of the life cycle activities. Planning may need to consider protection of proprietary data that goes outside a company or even export controlled data that goes outside the country. Outsourcing may require a sub-supplier management plan, such as the one described in Section 13-3, Supplier Oversight Plans and Procedures, of FAA Order 8110.49.[1]

Planning takes an overall project perspective of the life cycle, activities, plans, standards, procedures, and the resultant documents and data, and finally, the hardware itself. **Figure 3.4** shows this type of perspective.

DO-254 planning produces the plans, standards, and procedures for the project. The PHAC is the central output of the planning process as it starts the certification liaison activities with the certification authority and also invokes the plans and standards for the project. The central role of the PHAC is shown in **Figure 3.5**.

For DO-254, the plans should be written such that each plan is consistent within itself and also when all of the plans are considered as a group. Internal consistency means that terms used in the plan are used the same way, with the same meaning and the same spelling throughout the document. Internal consistency also means that references to sections, tables, and figures are correct. Consistency among a group of plans means that each plan uses terms in the same way, with the same meaning and with the same spelling. The transition criteria between the design plan, verification plan, validation plan, process assurance plan, and configuration management plan should all align. The outputs from a life cycle phase should serve as the inputs to the next life cycle phase. Outputs from the design process should serve as inputs to the verification and validation processes. Outputs from the design process activities should be the same data or documents used as inputs for the validation, verification, process assurance, and configuration management activities. Plans can start as hardware control category 2 (HC2) controlled while they are written. The PHAC is controlled as hardware control category 1 (HC1) after it has been released. The hardware configuration management plan is HC1 controlled for DAL A and B projects and can be HC2 controlled for DAL C and D projects. All other hardware management plans and standards can be HC2 controlled. In some cases, it may be beneficial to control plans and standards as HC1 data to ensure that changes to the plans or standards are reviewed, approved, released, and communicated to impacted parties.

DO-254 allows for various packaging of hardware life cycle data. Projects can combine plans and/or standards as long as the basic required content is provided. Note that embedding standards in a PHAC or combining other plans with the PHAC makes the included data HC1 controlled. Combining other plans or standards with the PHAC will also mean that the combined document is submitted to the FAA since the PHAC is a submittal document.

The plans produced in the planning phase and the respective DO-254 section describing the contents are:

- Plan for Hardware Aspects of Certification — DO-254 Section 10.1.1
- Hardware Design Plan — DO-254 Section 10.1.2
- Hardware Validation Plan — DO-254 Section 10.1.3
- Hardware Verification Plan — DO-254 Section 10.1.4
- Hardware Configuration Management Plan — DO-254 Section 10.1.5
- Hardware Process Assurance Plan — DO-254 Section 10.1.6

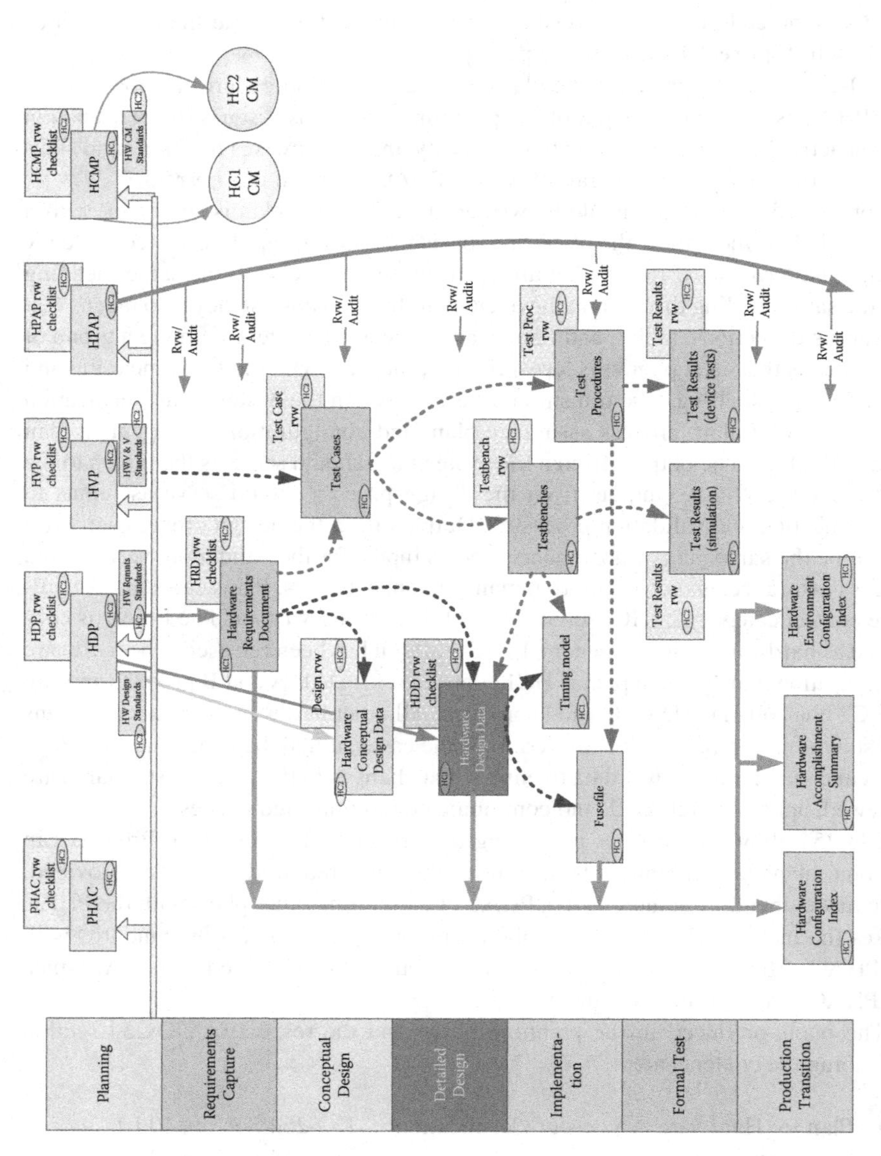

FIGURE 3.4 **(See Color Insert.)** DO-254 Life Cycle and Data

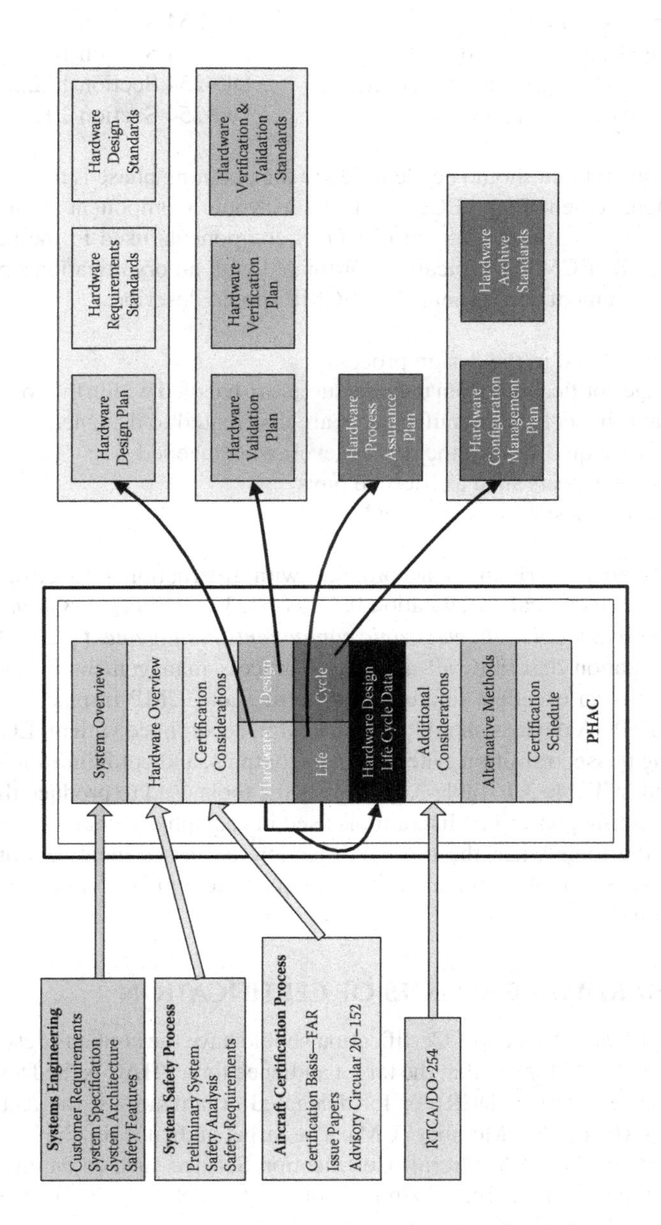

FIGURE 3.5 **(See Color Insert.)** Role of Plan for Hardware Aspects of Certification

The standards produced in the planning phase and the respective DO-254 section describing the contents are:

- Requirements Standards DO-254 Section 10.2.1
- Hardware Design Standards DO-254 Section 10.2.2
- Validation and Verification Standards DO-254 Section 10.2.3
- Hardware Archive Standards DO-254 Section 10.2.4

Another document that should be identified in the planning phase is the Electronic Component Management Plan (ECMP). The Electronic Component Management Plan defines the use and management of COTS components used in the hardware design process. The ECMP is typically coordinated with an organization's purchasing or parts procurement department. The ECMP should describe:

- The parts allowed in the design process
- How change notifications from manufacturers are handled within the company
- How errata sheets from manufacturers are distributed to designers
- How parts are qualified (at the device level) when needed
- Procurement aspects such as lifetime buys
- How parts obsolescence is managed

The ECMP can be written in accordance with International Electrotechnical Commission (IEC) technical specification IEC TS-62239 titled *Process management for avionics—Preparation of an electronic components management* plan.[2] The IEC technical specification describes all aspects of process management for electronic components and is an excellent reference for putting an ECMP in place. Note that unprogrammed FPGA devices should be managed in accordance with an ECMP.

The planning phase, transition criteria, inputs, outputs, and configuration controls are summarized in **Table 3.1**. Table 3.1 also lists the tools used to produce the documents in the planning phase. The list of tools used in each phase makes a convenient way to survey the tools when the tool qualification and assessment section of the PHAC is created. The table is meant as an example; a real project may require different information.

PLAN FOR HARDWARE ASPECTS OF CERTIFICATION

The Plan for Hardware Aspects of Certification should have the contents described in Section 10.1.1 of DO-254. Note that the target audience for a PHAC is the Designated Engineering Representative (DER), or for delegated organizations, the Authorized Representative (AR) or Unit Member (UM). The target audience for the PHAC also includes the staff of the FAA Aircraft Certification Service such as an engineer in the Aircraft Certification Office. It is important that the PHAC writer considers the target audience and readers of the document. In other words, the PHAC is written to satisfy the needs, questions, and criteria that the readers will use to evaluate the document. Since the PHAC is a submittal document, it is also important that the PHAC is written in a professional style and is free of inconsistencies and typographical

TABLE 3.1
Planning Phase Table

Phase	Entry Criteria	Activities	Design Tools	Output	CM Storage	Exit Criteria
Planning	• Certification basis defined • Issue Papers identified • FFPA prepared • PSSA prepared • FHA complete • System description available • Peer review comments available • PR written to update plans or standards	Prepare or update hardware management plans	• Word • Excel • Visio • PowerPoint	• PHAC • HDP • HVP (verification and validation) • HPAP • HCMP	• PHAC & HCMP HC1 controlled • HDP, HVP, HPAP HC2 controlled	Hardware management plans released
		Prepare or update hardware standards	• Word • Excel • Visio • PowerPoint	• HW Requirements Standards • HW Design Standards • Verification Standards • Archive Standards	Hardware standards HC2 controlled	Hardware standards released

errors. A review of the PHAC by an experienced technical writer can help ensure a professional finished document. In a formal sense, the PHAC is written in future tense, laying out the plan for the future planned hardware development.

The PHAC is an FAA submittal document—it is sent to the FAA with an approval or recommendation for approval. If a DER/AR/UM is involved in the program, they will have recommend or approval authority for the PHAC. If the FAA retains approval authority of the PHAC, then it is submitted as recommended for approval and the FAA will make the final document approval and concur with the recommendation for approval. If the DER/AR/UM has approval authority and the FAA does not retain compliance, then the PHAC can be approved by the designee.

The PHAC describes how DO-254 is interpreted and applied for approval of the airborne electronic hardware for a given program. A PHAC is an instantiation of DO-254, with program specific details. As the PHAC represents an agreement with the FAA on certification aspects for airborne electronic hardware, it is essential to get the PHAC written, approved, and submitted to the FAA as early as possible in a program. Securing FAA concurrence with a recommendation of PHAC approval, or FAA approval of the PHAC, is an important milestone. Once approved, the program is performed in accordance with the PHAC, not DO-254.

The PHAC typically starts with a description of the system. This description sets the context for the readers and helps them evaluate and understand the functions and safety aspects. The system description includes the architecture and functionality of the system, the failure conditions, and references to applicable system level documentation. The system description also states how functions are allocated to hardware and software.

Drilling down to the next level, the PHAC describes each hardware item in the scope of the PHAC. The item description should include the device types and technologies, such as flash, anti-fuse, or static random access memory (SRAM) devices. The device type should also list the manufacturer name and part number, and include a list of device features such as internal random access memory (RAM), hardware functions in the device fabric (multipliers, processor cores, etc.), and the overall device size or gate count. Any new or unique technologies used should be included. The hardware item's description should include safety aspects such as fail-safe, dissimilarity, partitions, and redundancy. Note that while circuits with discrete components may be readily shown to be of different design assurance levels and physically partitioned, it is not anticipated that claims of physical or functional partitioning within a PLD are used. Even with a device layout or floor planner option, PLDs use common routing resources and clocks, along with common power, ground, and physical package.

The certification considerations section of the PHAC should state the certification basis for the program. The certification basis is the applicable FARs and their respective amendment levels. Typically for a Part 25 program, this is 14 CFR Part 25.1301 and 25.1309. The amendment levels of the FARs can be put directly into the PHAC, or a reference to the system certification plan, which defines the amendment levels for all the applicable FARs, can be used. If the program has Issue Papers or similar items from foreign authorities, this is an ideal section to explain the proposed means of compliance to each Issue Paper. This section should also explain the proposed means of compliance to the FARs. The means of compliance should be stated as DO-254, in accordance with FAA Advisory Circular 20-152.

The design assurance level and functional failure path analysis should be summarized in the certification considerations section. The functional failure path analysis is used to show which circuits and/or components cause a failure. Since the failures are classified as hazards, the design assurance level associated with each failure is known. For PLDs, the DAL is determined by the most critical hazard associated with the device. For LRU application of DO-254, the functional failures and DAL of each circuit can be established. With proper partitioning and isolation it is possible to have multiple DALs for electronic hardware (non-PLD). An effective approach is to list the top-level events from the functional hazard assessment for the hardware. Then, describe the functional failure path(s) and the device(s) or electronic hardware in each functional failure path. Once the hazards, failures, and hardware are associated, the design assurance level is then stated. If any reduction of design assurance levels is permitted due to architectural aspects, then the redundancy/dissimilarity/control-monitor aspects should be stated along with a justification of the resultant design assurance level.

Additional topics to cover in the certification considerations section are a list of the submittal documents to the FAA for the program, the designee, and their authority requested, and the list of SOI audits used in the program.

The hardware design life cycle section of the PHAC includes an overview of the development methods and the development procedures. This section also describes the procedures and standards that will be applied. The design life cycle is typically described by phases. Each phase has transition criteria that consist of entry and/or exit criteria. The transition criteria describe the activities that need to be completed and the associated artifact or evidence produced that demonstrates the completed activity. Each life cycle phase should also list the tools used for the respective activities. This section of the PHAC is meant to be an overview of the design life cycle, thus the specific plans and standards should be referenced. Table 3.1 shows a concise method to collect and portray the entry criteria, exit criteria, activity, tools, outputs, and configuration controls for a life cycle phase. It is also helpful to reference the DO-254 objective(s) associated with the life cycle phase or a specific activity. This will be useful to the certification authority that reads and needs to approve the PHAC.

The hardware design life cycle data section of the PHAC explains how the program will deliver the data specified in DO-254 Table A-1. It is recommended that a table be included in the PHAC that shows the mapping between the life cycle data listed in DO-254 Table A-1 and the project specific life cycle data. When documents are combined, they can be listed multiple times in the PHAC table. An example of the life cycle data mapping for an FPGA is shown in **Table 3.2**. The document identifier should use the project specific designation. When DO-254 is used at the system level, the data should be more similar to what is shown in DO-254. In other words, the Top-Level Drawing would actually be used and not replaced with the PLD equivalent document.

The additional considerations section of the PHAC includes a description of any previously developed hardware, usage of COTS components, product service experience, and an assessment of the development and verification tools used in the hardware life cycle. Previously developed hardware (PDH) can be claimed when the hardware, or PLD device, was developed and approved prior to FAA Advisory

TABLE 3.2
Hardware Life Cycle Data

DO-254 Document	Company Document	Document Identifier	FAA Submittal	DER Disposition
Plan for Hardware Aspects of Certification	Plan for Hardware Aspects of Certification X123 PLD	1002-PHAC	Yes	Recommend
Hardware Design Plan	FPGA Design Plan	1002-HDP	No	Approve
Hardware Verification Plan	FPGA Verification Plan	1002-HVP	Yes	Approve
Hardware Validation Plan	FPGA Verification Plan	1002-HVP		
Hardware Process Assurance Plan	FPGA Process Assurance Plan	1002-HPAP	No	Approve
Hardware Configuration Management Plan	FPGA Configuration Management Plan	1002-HCMP	No	Approve
Requirements Standards	FPGA and Hardware Requirements Standards	1002-HRSTD	No	Approve
HDL Coding Standards	VHDL Design Standards	1002-VHDSTD	No	Approve
PLD Design Assurance Standards	PLD Design Standards	1002-PDS	No	Approve
Validation & Verification Standards	PLD Verification Standards	1002-VVSTD	No	Approve
Hardware Archive Standards	Archive, Retrieval, and Data Backup	1002-HWARSTD	No	Approve
Hardware Requirements	X123 PLD Requirements Document	1002-HRD	No	Approve
Conceptual Design Data	X123 PLD Design Document	1002-HDD	No	Approve
Detailed Design Data	X123 PLD Design	1002-HDL	No	Approve

TABLE 3.2 (continued)
Hardware Life Cycle Data

DO-254 Document	Company Document	Document Identifier	FAA Submittal	DER Disposition
Top-Level Drawing	X123 PLD Hardware Configuration Index	1002-HCI	Yes	Recommend
Assembly Drawing	X123 Netlist	1002-NET	No	Approve
Installation Control Drawing	X123 Altered Item Drawing	1002-AID	No	Approve
HW/SW Interface Data		n/a		
Hardware Traceability Data	X123 PLD Hardware Verification Report	1002-HVR	No	Approve
Hardware Review & Analysis Procedures	X123 PLD Hardware Verification Procedures	1002-HVP	No	Approve
Hardware Review & Analysis Results	X123 PLD Hardware Verification Report	1002-HVR	No	Approve
Hardware Test Procedures	X123 PLD Hardware Verification Procedures	1002-HVP	No	Approve
Hardware Test Results *includes elemental analysis results*	X123 PLD Hardware Verification Report	1002-HVR	No	Approve
Hardware Acceptance Test Criteria		n/a		
Hardware Environment Configuration Index (Order 8110.105[a])	X123 PLD Hardware Environment Configuration Index	1002-HCI	No	Recommend
Hardware Accomplishment Summary	Hardware Accomplishment Summary X123 PLD	1002-HAS		

[a] Order 8110.105 CHG 1, *Simple and Complex Hardware Approval Guidance*, Federal Aviation Administration, dated September 23, 2008.[3]

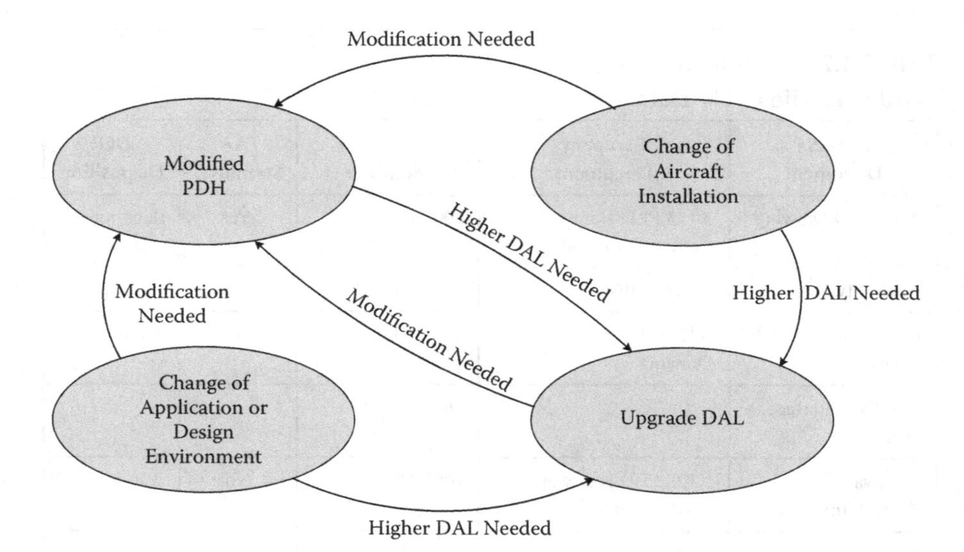

FIGURE 3.6 Decision Process Flow for Previously Developed Hardware

Circular 20-152. The description of previously developed hardware should state the FAA project number that the hardware was originally approved under, the DAL for the original approval, the life cycle data substantiating the approval—such as the PHAC, HCI, and HAS, and summarize any applicable product service experience.

In addition, the previously developed hardware section should address whether the hardware is being reused as is, with an upgrade to the DAL, with a change to the hardware, with a change to the aircraft installation, with a change to the application or design environment, or with some combination of these considerations. **Figure 3.6** illustrates the decision process flow for PDH.

The amount of reuseable data for PDH will depend on project circumstances. Unmodified PDH will only require that integration/system testing be conducted. Modified PDH will require:

- Test and verification of the modified aspects of the hardware
- Retest and reverification of parts that are affected by modified parts
- Integration/system testing

Changing the application will require:

- Reverification of interfaces to new equipment
- Reverification of hardware/software interface if different software used

A change of the design environment requires an assessment of the new design and/or verification tools for tool qualification.

An increase of the design assurance level requires:

- Assessing suitability of previous compliance data for new DAL
- Filling in gaps between previous and new DAL
- Determining which existing compliance data can be used

Commercial off-the-shelf components, including unprogrammed FPGAs, should be managed in accordance with an ECMP. Other COTS aspects should include the use of intellectual property (IP) cores. IP cores are designs that are purchased from a vendor that may or may not have been developed for DO-254 compliance. IP cores are typically used for standard interfaces such as Ethernet, 1553 bus, ARINC-429, universal asynchronous receiver/transmitter, or peripheral component interconnect (PCI). The life cycle data for an IP core should be commensurate with the DAL of the hardware. In other words, the life cycle data and verification activities should be the same as designs conducted in-house. When the IP core does not have the HDL source code available, it could preclude its use. The PHAC should describe the IP core, the documentation available from the vendor, and explain any additional activities that need to be performed to complete the life cycle activities and data. Structuring the IP core data as standalone documents is an effective strategy, especially when the IP core will be reused in subsequent designs.

Product service experience is used to justify the use of COTS components and previously developed hardware. The product service experience would ideally come from aircraft use. If aircraft usage data is not available, then an explanation should be included to justify the relevance of the data. There is always a first time for a device to be used in an aircraft application; the idea is to convey whether the device has been used in other industries or applications for a sufficient amount of time. Semiconductors that have been used in automotive applications are often ideal candidates since the device packaging and temperature ranges are often suitable for aerospace applications. The product service experience should explain why the data is meaningful and relevant. The data should explain:

- Whether the function and usage is the same or similar
- Whether the operational environment is the same or similar
- Whether the previous DAL is the same or similar to the proposed DAL
- Whether the hardware configuration is the same or similar
- Any errors or problems detected during the service period
- Failure rates during the service period

Product service experience can also be used to justify the upgrade of the design assurance level from D to C, C to B, or B to A. The service hours should be on the same order of magnitude as the required reliability. When assessing service hours for DAL A, the prior usage should be in the millions of hours since the failure rates will be on the order of 10^{-5} to 10^{-6}.

TOOL ASSESSMENT AND QUALIFICATION

DO-254 requires that the tools used in hardware design and verification processes be assessed. The flow chart in Figure 11-1 of DO-254 illustrates the tool assessment and qualification process that must be conducted on each design and verification tool.

Tools can be classified as design tools or verification tools, or in some cases both. Design tools are the tools that are used to generate the hardware or some aspect of its design. This means that an error generated by a design tool can introduce errors into the electronic hardware. Examples of PLD design tools include synthesis and place and route tools. For electronic hardware such as circuit cards, examples of design tools would include schematic design entry tools and circuit board layout software. Verification tools have less effect on design assurance than design tools because their most serious effect is to fail to detect an error in the electronic hardware or its design. Examples of PLD verification tools include simulation tools, logic analyzers, and oscilloscopes. For electronic hardware, verification tools include analog and mixed-signal circuit simulation software, laboratory test fixtures, and bench instruments such as logic analyzers and oscilloscopes.

DO-254 does not provide specific rules or guidance for qualifying design tools. Design tool qualification can follow strategies described in DO-254 Appendix B, or they may use the guidance for qualification of software development tools provided in RTCA DO-178. Regardless of the approach, design tool qualification should be described in the PHAC and coordinated in advance with the certification authorities. In addition, design tool qualification will require cooperation from tool designers and ample schedule and budget resources.

The emphasis in DO-254 is on verifying a tool's output. If the output of a design or verification tool is independently assessed, then no tool qualification is needed. Independent assessment includes any process that verifies that the tool output is correct, including the use of another tool to make the assessment, and may include a manual review of the tool output or the use of a separate, but dissimilar, tool to conduct a similar check. Qualification is not required for verification tools used for measuring completion of elemental analysis. DO-254 also allows the use of relevant tool history as a substitute for qualification.

Design tool qualification requires the generation of a tool qualification plan. The rigor of the qualification effort is determined by the type of tool, its intended application, and the design assurance level of the hardware that it will design. Compliance with the tool qualification plan is documented in a tool accomplishment summary.

If the output of the tool is not independently assessed, and there is no relevant history, Level A, B, and C design tools and Level A and B verification tools require a "basic" tool qualification. The "basic" tool qualification verifies that the tool produces the correct outputs for its intended application. The basic qualification tests the tool's operation against its "requirements," which are typically taken from its user manual or equivalent documentation of its functionality.

Tool assessment and qualification for each tool, including any rationale for qualifying or not qualifying the tools, should be documented in the PHAC. If relevant tool history will be used for a design or verification tool in lieu of assessing its outputs, or if a design tool will be qualified, they should be discussed in the PHAC

TABLE 3.3
Tool Assessment Example

Vendor	Tool	Purpose	Assessment	Rationale
Syn Free Corp.	SynCity	HDL synthesis tool	Qualification not required	Functional and post-layout simulation and device testing performed with verification tools
MisPlaced Tools Inc.	Route 66	Place and route tool	Qualification not required	Post-layout simulation and device testing performed with verification tools
ProLogicDesign Associates	Write Way	HDL editor	Qualification not required	HDL text files output from the tool are reviewed
Runner Up Inc.	Run For Cover	HDL code coverage	Qualification not required	Not required

and coordinated with the certification authorities during the planning phase of the program.

Tool assessment in the PHAC should list each design tool and verification tool. Adding the purpose of the tool and a brief description will help convey the usage and context. The PHAC should state whether the tool needs to be qualified, how the qualification will be performed, and the data produced as evidence of the qualification. If qualification is not required, the PHAC can state the rationale for not qualifying the tool. An example of the assessment can be formatted as shown in **Table 3.3**. The tools listed in the example are fictional.

ALTERNATIVE METHODS

The alternative methods section of the PHAC includes a description and justification for any methods that are not discussed in DO-254 or methods that are not applied as described in DO-254. Alternative methods might include use of development, verification, or validation techniques that are less traditional than those described in DO-254. The use of any alternative methods should be coordinated and discussed early in the program to minimize risk associated with the approach.

SCHEDULE

The certification schedule section of the PHAC describes the major milestones for the program and the timeline for submitting life cycle data to the certification authority. This information allows the FAA to plan their workload and allocate resources for the program. The major milestones may include dates or timeframes for first flight, type inspection authorization (TIA), equipment or aircraft certification, and the intended SOI audits. Note that dates should be kept to a high level of granularity since specific target dates can be subject to change. The intent is to convey a timeframe such as 4Q2014 (fourth quarter of 2014) rather than November 12, 2014.

FAA ORDER 8110.105 ASPECTS

The PHAC should also contain information specified in FAA Order 8110.105. The additional content for a PHAC from Order 8110.105 includes:

- A PHAC can be written for each hardware item or PLD in the equipment, or the PHAC can encompass all the PLDs in the equipment. Using separate PHACs for each PLD can be useful if the intent is to develop a PLD for reuse on future programs.
- The PHAC should state whether the hardware item is classified as simple or complex, and provide a justification if the classification is simple. Devices or hardware classified as simple should have a description of the proposed approach including the life cycle activities, data, and applicable plans and standards. The description for simple hardware should also describe the verification activities and environment and give a clear description of how the hardware is comprehensively tested.
- The PHAC should list the failure condition classifications associated with each hardware item. The PHAC should also provide a functional description of each hardware item.
- While DO-254 Sections 9.1 and 10.1.1.3 state that the proposed means of compliance should be stated in the PHAC, Order 8110.105 reiterates this point. Typically, the proposed means of compliance in the PHAC simply states that the hardware will be developed to meet the objectives in DO-254 commensurate with the design assurance level in accordance with FAA Advisory Circular 20-152.
- The Order also requires a description of the proposed design assurance level, and justification (this duplicates DO-254 10.1.1.3).
- The PHAC should also reference the applicable hardware management plans and hardware design standards. A list of certification data to be delivered and/or to be made available to the FAA should be in the PHAC (this duplicates DO-254 10.1.1.5).
- Any proposed alternative methods to those in DO-254 should be explained. The explanation of alternative methods should state how the objectives and guidelines are interpreted, and provide a description of the actual alternative methods. Alternate methods are also required per DO-254 Section 10.1.1.7. It is important to note that the FAA anticipates that usage of alternate methods is presented along with the compliance justification early in the project.
- A justification for reverse engineering life cycle data and activities for an existing component.
- Documentation and justification of relevant service history used to assess and preclude the qualification of tools. This should describe the version of the tool used on previous projects, any anomalies noted with the tools, any problem reports recorded for the tool, the way the tool has been used

on other projects, and whether the tool output/data/reports are accurate and reliable.
- A description and justification of the level of verification coverage of the requirements that will be achieved by hardware testing. Note that hardware testing assumes tests performed on flight hardware (i.e., in circuit testing with a production equivalent circuit card). Some portion of the hardware testing can also be performed with a device tester as long as in-circuit tests at the board level are also performed. The PHAC can describe the approach if a combination of device tests and in-circuit tests are used.
- The completion criteria for the additional verification activity selected from DO-254 Appendix B.

Content for the PHAC not specifically listed in DO-254 that is particularly useful includes:

- The LRU or equipment part number
- The part number of the end hardware item (PLD)
- The part number of the HDL source code
- The part number of the netlist, fusefile, or programming file applied to the unprogrammed part
- The vendor name and part number of the unprogrammed PLD part
- The FAA certification project number
- A list of applicable FAA Issue Papers and a detailed description of the proposed means of compliance to each aspect of the Issue Paper
- A list of applicable foreign certification authority items (such as an EASA Certification Review Item) and a detailed description of the proposed means of compliance to each aspect of the item

HARDWARE DESIGN PLAN

The Hardware Design Plan is for the development engineers to guide the design of the hardware. The hardware design life cycle section of the HDP describes the processes to create the life cycle data and the hardware item. The hardware item could be a system, a circuit card assembly, or a programmable logic device. The design life cycle should also explain the coordination between the hardware design, process assurance, configuration management, verification, validation, certification liaison, and production transition.

The coordination between these various life cycle processes is expressed as transition criteria. Transition criteria can specify the minimum requirements in order to start a life cycle phase or process, the minimum requirements in order to finish a life cycle phase or process, or both. While the criteria can state an activity that needs to be complete, the criteria can also list the artifact or evidence that demonstrates that the activity is complete. For example, the transition into conceptual design can formally start when the requirements are baselined and have been peer reviewed. The evidence that the baseline occurred and that the peer review was performed would be

TABLE 3.4
Transition Criteria

Phase	Entry Criteria	Activities	Design Tools	Output	CM Storage	Exit Criteria

a requirements document under configuration control and a completed peer review checklist stored in configuration control. As was shown in previous examples, the activities, associated artifacts, and transition criteria can readily be expressed in a table. **Table 3.4** shows an example of the format.

Table 3.5 shows an example of using a table to describe the requirements capture process.

The hardware product description section of the HDP should include the intended usage of the hardware including the operational environment, any alternate usage, the target service life, and any plans for updates or upgrades to address obsolescence or features. The description should include device programming considerations for programmable devices such as in-circuit or device programmer. If the HDP is not specific to a project, the hardware product description can be captured in the project PHAC, and the hardware product description section of the HDP can then refer to the project PHAC.

The hardware design methods section of the HDP should include a description of the processes and procedures for capturing requirements, performing the conceptual design, performing the detailed design, and transferring the necessary data to production. Information on the use of requirements capture or management tools and tools for design or schematic capture should be included or referenced from the design plan. The design plan should also explain the data needed for the production environment such as fuse file, programming file, Gerber file, parts list, bill of materials, assembly drawings, and programming instructions for programmable devices. The design plan should also specify the types of parts permitted in a design and reference the Electronic Component Management Plan for parts selection criteria.

The design plan should explain the intended levels of requirements such as system, assembly, board, and PLD. The requirements method should explain how requirements are captured, the use of requirements management tools, how derived requirements are captured, and how the trace data for requirements is captured.

The hardware design environment section of the HDP should list the tools used in the design process. The tools should include software packages, hardware or computing platforms, and any specialized equipment.

The hardware item data section of the HDP provides a complete identification of the hardware to be produced. This should include the part number of the end system, the part number of sub-assemblies, and the part number of programmable devices. For programmable devices, the identification should include the manufacturer's part number, the part number of the programming data, and the resultant part number of the programmed device. If previously developed hardware is being used, then this section should list the complete configuration identification of the hardware item and

TABLE 3.5
Transition Criteria for Requirements Capture

Phase	Entry Criteria	Activities	Design Tools	Output	CM Storage	Exit Criteria
Requirements Capture	• HDP released • HW Requirements Standards released and approved • Board level requirements released • Requirements allocation to AEH complete • Board design started • Peer review comments available • Problem report (PR) written to update requirements	• Define functional elements • Create or update PLD requirements from functions allocated to PLDs • Create or update trace links from PLD requirements to parent requirements • Create or update any additional derived requirements • Add or update justification for derived requirements • Create or update trace links for derived requirements	Word, Excel, Visio, DOORS	• DOORS baseline export snapshot • Cut set of trace data • PLD Hardware Requirements Document	• DOORS baseline export snapshot HC2 controlled • Cut set of trace data HC2 controlled • Hardware Requirements Document HC2 controlled	• PLD Hardware Requirements Document HC1 controlled • PLD Hardware Requirements Document released

life cycle data that is being reused. The description should also clearly identify the new or updated hardware and life cycle data.

The other considerations section of the HDP includes a description of any factors needed for the production of the hardware including mounting, installation, programming, pick and place equipment limitations, solder reflow profiles, and provisions for manufacturing test.

HARDWARE VALIDATION PLAN

The Hardware Validation Plan describes the processes and activities to validate the derived hardware requirements. DO-254 processes assume that all requirements allocated to the system have been validated in accordance with ARP4754. Thus, only the derived requirements added by the hardware design process need to be validated. The validation of derived requirements should satisfy the objectives in DO-254 Section 6.1.1. In most cases, the validation process for PLD derived requirements can be included in the Hardware Verification Plan. Derived requirements for a PLD typically involve a requirements review to check that the PLD derived requirements are correct and complete. Validation of derived requirements for application of DO-254 to all electronics in a system may involve activities beyond review. All requirements, including validated derived requirements, are subject to verification as described in the Hardware Verification Plan.

The validation methods section of the Hardware Validation Plan includes a description of how reviews are performed to validate derived requirements, how analyses are performed to validate derived requirements, and how testing is performed to validate derived requirements. The analysis may include simulation methods.

The validation data section of the Hardware Validation Plan describes the data produced from the selected validation method. For review of derived requirements, this can be a completed requirements review checklist. For analysis or simulation of derived requirements, this can be analysis or simulation results. For testing of derived requirements, this can be test results. The validation environment section of the Hardware Validation Plan describes the tools, checklists, and equipment used to validate derived requirements.

HARDWARE VERIFICATION PLAN

The Hardware Verification Plan describes the processes and activities to verify the hardware requirements. The requirements verification should satisfy the objectives in DO-254 Section 6.2.1. The verification methods section of the Hardware Verification Plan describes the various reviews, analyses, and testing performed and the methods used to accomplish them. The descriptions of the methods should include the methods' processes, methodologies, activities, input data, and output data. Helpful information in this section would include a summary of how the verification method will be selected for each requirement, the order of precedence among the methods, and the conditions that can dictate when more than one method should be used to verify a requirement.

The verification data section of the Hardware Verification Plan describes the data produced as a result of performing the verification activities. The data would include peer review records, analysis procedures and results, and test procedures and results. Helpful information for this section includes how the data will be managed and archived, how it will be analyzed or otherwise evaluated to generate the verification results, and how the data and results will be documented as the hardware verification results life cycle data.

When DAL A or B hardware is produced, the verification independence section of the Hardware Verification Plan specifies how the independence is achieved. This section describes how the verification will meet the independence described in Appendix A of DO-254. In general, independence is achieved when the reviewer of a verification artifact is different from the author of the artifact. Qualified or properly assessed verification tools also provide an independent assessment of verification data. Test cases should be reviewed by someone other than the author of the test cases. Test procedures should be reviewed by someone other than the author of the test procedures. Test results should be reviewed by someone other than the person running the tests and compiling the results. Analysis procedures and results should be reviewed by someone other than the author of the data. Note that DO-254 permits the use of verification cases, procedures, and results developed by the hardware designer as long as the reviewer is a different person. Use of verification data from the designer is also permitted when there is additional independent verification of the same requirement and design, such as in other levels of system or software testing—provided that they were developed independently.

Note that independence is not dictated solely by whether the reviewer is a different person than the author or originator of the data being reviewed. It is often the case that a person who did not contribute to the data being reviewed, but who is familiar with the data and the engineering thought processes that went into its creation, will be incapable of conducting an independent review simply because they are too close to the data to remain objective or to see the data with a fresh perspective. While such a person will technically satisfy the independence criteria in DO-254, in reality they may not possess the independence to conduct a faithful review as intended by DO-254. On the other hand, while complete ignorance of the data and its associated technology will ensure an independent review, it may result in an inadequate review simply because the reviewer will not know enough about the topic to recognize an anomaly or discrepancy. Thus a reviewer will need to have some level of familiarity with the data being reviewed to ensure that the review is competent, but should not be so familiar that their ability to identify problems—and therefore their independence—will be compromised.

The verification environment section of the Hardware Verification Plan should describe the tools and equipment needed to perform the reviews, analyses, and tests. Analysis tools would include the simulation tools for circuit analysis, thermal analysis, or PLD timing. Hardware test equipment should include power supplies, function generators, digital multi-meters, oscilloscopes, logic analyzers, test stands, break out boxes, standard or specialized cables, or interconnecting wiring. Plans should include provisions to make diagrams of test equipment setup and drawings for any custom test equipment or wiring. A table can be used to list the review checklists used in the review activities.

The organizational responsibilities section of the Hardware Verification Plan should describe the various organizations involved in the verification and the activities that they perform. If credit for hardware verification is taken from software testing, then the software organization should also be listed in this section. If credit for hardware verification is taken from system testing, then the systems organization should also be listed in this section. If any of the activities are outsourced to another company or organization, then this section should specify those details. Things to consider and specify in the Hardware Verification Plan when outsourcing verification activities include:

- Technical oversight of the outsourced work—the amount of technical review performed to assess the quality and correctness of the data
- Impact on certification liaison, such as whether an audit needs to be performed at another location
- Process assurance oversight of the outsourced work—which organization will perform process assurance activities
- Coordination of configuration management aspects
  - Where the data is managed
  - How compliance data is transferred back to the developer's configuration management system
  - How problems discovered by the verification team are transferred back to the developer's configuration management system
  - Which configuration management tools are used to control compliance data
- Whether tests or simulations will be run for score at the outsource facility. This can introduce complexity and additional cost to coordinate any test witnessing or audit aspects.

HARDWARE CONFIGURATION MANAGEMENT PLAN

The Hardware Configuration Management Plan (HCMP) can be thought of as a collection of procedures for the control of the hardware and its associated life cycle data. The HCMP defines the controls and procedures for two categories of configuration management for life cycle data. Hardware control category 1, or HC1, uses formal release, baselines, and problem reporting to manage changes. Hardware control category 2, or HC2, is used to store and protect life cycle data such as verification results or process assurance records. HC2 data does not require formal baselines, problem reporting, and change control; the data is simply archived and a new version created. HC1 data requires that all configuration management activities described in DO-254 be applied, while HC2 data requires a subset of those activities.

Table A-1 in DO-254 shows which hardware control category applies to hardware life cycle data. In general, the following data items are always HC1 controlled, regardless of the design assurance level:

- Plan for Hardware Aspects of Certification
- Hardware Requirements Data
- Top-Level Drawing (Hardware Configuration Index for a PLD)
- Assembly Drawing
- Installation Control Drawing
- Hardware/Software Interface Data
- Hardware Accomplishment Summary

The process activities associated with hardware configuration management include:

- Configuration identification
- Baselines
- Baseline traceability
- Problem reporting
- Change control integrity and identification aspects
- Change control records, approvals, and traceability
- Release
- Retrieval
- Data retention
- Protection against unauthorized changes
- Media selection, refresh, and duplication

Configuration management, especially baseline and data retention activities, should be applied to the hardware development and verification tools. Commercially procured tools and in-house custom made tools should be preserved for any necessary future rework of the hardware item.

Configuration identification pertains to providing a unique identifier to each configuration item. All life cycle data such as drawings, documents, verification artifacts, process assurance records, and configuration management records should have a method to assign unique identifiers. File management tools with version control will ensure that changes to life cycle data have unique identifiers since the version is rolled with each check in of the data in the file management tool. Companies typically have part numbering schemes established through drawing room policies and procedures. Logs can be kept manually or electronically that track which drawing, document, or hardware identifier numbers have been allocated. While it is obvious that hardware and drawings should have unique part numbering schemes, DO-254 requires that all HC1 and HC2 life cycle data be uniquely identified. Many PLD designs could have a "`top.vhd`" or "`arinc.vhd`" file. Ensuring that the files are unique would require that the identifier include the entire file path name specification, and that the file path is unique for each project. An example would be `C:\serverroot\projects\projectAlpha\designfiles\source\design`, where `projectAlpha` is the unique project name. The file "`top.vhd`" could then be distinguished between

- C:\serverroot\projects\project**Alpha**\designfiles\source\design\top.vhd
- C:\serverroot\projects\project**Beta**\designfiles\source\design\top.vhd

The unique identifiers need to be applied to all verification data produced during the life cycle. This includes data for physical hardware testing—test cases, test procedures, and test results—and to data for hardware simulation—test cases, testbenches, and test results. Peer review records from requirements, design, and verification data reviews also need to be uniquely identified. A structured naming practice that starts with the requirements can make it relatively easy to meet the unique identification constraint. Structured naming practices coupled with a structured hierarchy for file organization make storage and retrieval of data across multiple projects a manageable feat.

Baselines apply to individual documents, drawings, and data as well as collections or sets of data. A baseline can refer to the first release of a document, typically called *Revision (—)* or *Revision 0*, or it can refer to a collection such as the planning baseline. The planning baseline would be the collection of released hardware management plans and hardware standards and their respective revision level. Baselines can have maturity criteria for entry into system level environmental qualification testing, aircraft ground test, or aircraft flight test. For a PLD, a baseline for a particular version of the hardware is described in the hardware configuration index, or HCI.

Baseline traceability allows a description of the changes incorporated between subsequent releases of HC1 data. For a document, this would be the problem reports incorporated in a version to create the subsequent version, such as changes to Rev. A of a document to create Rev. B. Baseline traceability also allows a description of the changes incorporated between subsequent releases of the hardware item. If the top-level part number uses a XXXXX-YYY numbering scheme where XXXXX is the root number and YYY is the version, the 13579-001 can be traced to 13579-002 through the problem reports that describe the changes applied to the –001 version to create the –002 version. Systems and LRUs use a top-level drawing for the hardware to call out the drawings and their respective version. Baselines are particularly useful for accumulating credit for the verification activities when changes are being made to the hardware. Skillful use of baselines, problem reporting, change control, and change impact analysis can help justify where verification credit can be retained, and identify which verification activities must be repeated when a change is made.

Problem reporting uses a process to report, track, and disposition the status and resolution of problems. Often, the problem reporting system is also used for change management to incorporate new, changed, or even deleted functions and requirements. Problem reports are most effectively managed using electronic or web-based tools that also enforce the work flow and sign off of problem reports (PR). Problem reporting tools facilitate work flow automation, notification to users, and reporting for project management.

Change control starts when a configuration item is baselined and used for certification credit. Typically for HC1 life cycle data, this starts when a document or data has been reviewed and released. Once the life cycle data is baselined, protections should be put in place and enforced to ensure that any subsequent change is an authorized change. Since change control is usually linked with problem reporting, this usually means that life cycle data require a PR and approval from the change board to make changes once the data has been released. Drawings may follow a similar change and approval process, or use the problem reporting system. Any change to baselined or released life cycle data should cause a change to the configuration identification. This is typically done by incrementing the revision or version of the data or document. A document initially released as Revision N/C (no change) would be incremented to Revision A with the first authorized change(s). A testbench `vhdl` file would increment from version –001 to –002, a schematic would increment from Rev. A to Rev. B, and so on. The type of increment will be compatible with the storage tools and methods. Tools used for file management, such as Rational ClearCase, CVS, MKS, AccuRev, and others will automatically increment the file version for each check in. If these tools are used for management of PLD design and verification data, then their versioning information can be used to ensure the integrity and identification aspects of change control. Manual methods to ensure the integrity and identification aspects of change control, while more time consuming, are also acceptable.

The coordination of problem reporting and version control should be described so that changes made to HC1 controlled released data are recorded, approved, and traceable. Once the change process starts, the PR should describe the starting version of the life cycle data and the resultant version when the changes are made. Document version updates and version updates to HC1 data in file management systems should be described. Updating the version of changed items facilitates tracing changes back to the original, or former, version. Problem reports should also record the changes made and the approval to make the changes.

The HCMP should describe the procedures used to release data, drawings, and documents. The release process is used to ensure that the baselined and controlled version of data is used in subsequent activities. An example is a released requirements document that is used for creating the design and the trace data from requirements to the design. The release is a promotion of the document or data from engineering control to a formal configuration management system. Released documents or data can be stored in access-controlled areas of a server, in a file management system, or a company production data management system. Any forms, approvals, routing, or controls should be explained in the procedure.

The HCMP also describes how life cycle data is retrieved. The retrieval can be from a server or application such as a file management system or from a formal company data management system. Data retrieval from off-site archives should also be described. It is prudent to periodically retrieve data from off-site archives and practice restoring databases, applications, and servers. Documenting the detailed procedures to retrieve and restore data will prove useful in the event of data loss or corruption.

Project life cycle data should be retained as long as the equipment is used in service on an aircraft. The data for FAA projects should be retained in the United States and written in English. Many corporate business systems or information technology services are set up to accommodate seven years data retention for legal or business purposes. The FAA requirement could easily extend the retention period to 20 or 30 years, perhaps longer. In addition to the life cycle data, it is important to retain the design and verification tools and a license to allow their use.

All released and archived data should be protected against unauthorized changes. The HCMP should describe server access controls, user account controls, and other methods used to ensure that the data is protected. Most file management tools facilitate this protection by updating the version of data or documents when they are checked in. The original version can always be accessed. Production data systems typically only allow read access to stored data; administrators are the only ones allowed to add data or documents. Problem reports for HC1 data record authorization to update or change released data and documents.

The HCMP should also describe how the media used to archive project data is selected, and when it is refreshed or duplicated. Optical, electronic, and magnetic media should use multiple copies in case of damage to or degradation of the media. Maintenance of archives should be described with considerations for the type of media. Periodic assessments of data storage technologies can be performed to evaluate more cost-effective or efficient technologies. Changes in media should be accompanied by data migration from old media to new media. It is very useful to describe the data produced as evidence of the management of project storage media. This will allow process assurance to audit records.

HARDWARE PROCESS ASSURANCE PLAN

The Hardware Process Assurance Plan describes the activities performed by process assurance. Process assurance needs to be performed independently of other life cycle design and verification activities. If a separate organization is not used for process assurance, then the plan should describe how the individuals maintain their independence from the life cycle tasks they were involved in when they perform process assurance activities. While DO-254 mentions product conformance in Section 10.1.6, the conformance is typically handled at the system level through formal FAA activities. Most projects describe conformity and delegation of conformity in the system level certification plan to ensure coordination with and delegation from the FAA. DO-254 applied at the system level or PLD level will use company configuration management processes, in lieu of formal conformity, for verification activities that meet DO-254 objectives. In this case, process assurance should ensure that test equipment is documented and that testing uses data from configuration controlled sources. The HPAP should describe the reviews and audits performed by process assurance. Process assurance should ensure that design and verification life cycle transition criteria are satisfied, that project standards are adhered to, and that all configuration management objectives are met.

The process assurance activities can be organized around the phases defined in the Hardware Design Plan and the Hardware Verification Plan. An example is shown in **Table 3.6**.

The table defines when a process assurance activity is performed, the specific activity needed, and the data produced as evidence of the activity. Note that process assurance does not necessarily need to be performed by technically oriented staff. Process assurance audits of the peer review process could be an independent peer review performed by process assurance, or process assurance could check that the peer review was performed correctly by engineering and that all issues captured in the review were resolved correctly.

Process assurance should define the level of auditing they will perform. For design and especially verification activities with multiple aspects, a sample target or

TABLE 3.6
Process Assurance Activities

Phase	Entry Criteria	Activities	HPA Tools	Output	CM Storage	Exit Criteria
Planning	• PHAC, HDP, HVP, HCMP, HPAP released • Hardware standards released	Complete the HPA Planning Completion Checklist	Word, Excel, Visio	• Completed Planning Completion Checklist	Planning Completion Checklist HC2 controlled	Planning updates completed
Requirements Capture	Hardware Requirements Document released	Complete the HPA Requirements Checklist	Word, Excel, Visio	• Completed Requirements Checklist	Requirements Checklist HC2 controlled	Requirements updates completed
Conceptual Design	Conceptual design of all functional elements complete	Complete the HPA Conceptual Design Checklist	Word, Excel, Visio	• Completed Conceptual Design Checklist	Conceptual Design Checklist HC2 controlled	Conceptual design completed
Detailed Design	Hardware Design Data HC2 controlled	Complete the HPA Detailed Design Checklist	Word, Excel, Visio	• Completed Detailed Design Checklist	Detailed Design Checklist HC2 controlled	Detailed design completed
ALL	Ongoing periodic audits	Audit PRs, CCB, peer review records, CM records	Word, Excel, Visio	Completed Periodic Audit Checklist	Periodic Audit Checklist HC2 controlled	

continued

TABLE 3.6 (continued)
Process Assurance Activities

Phase	Entry Criteria	Activities	HPA Tools	Output	CM Storage	Exit Criteria
ALL	Deviation detected	Record deviation and corrective action	Word, Excel, Visio	Completed Deviation and Corrective Action Worksheet	Deviation and Corrective Action Worksheet HC2 controlled	
Implementation	• Hardware Configuration Index released • Hardware Life cycle Environment Configuration Index released • Fusefile released	Complete the HPA Implementation Checklist	Word, Excel, Visio	• Completed Implementation Checklist	Implementation Checklist HC2 controlled	Implementation completed
Formal Test	• Hardware Test Procedures released • Programmed device • Test equipment set up	Complete the HPA Formal Test Checklist	Word, Excel, Visio	• Completed Formal Test Checklist	Formal Test Checklist HC2 controlled	Formal Test completed
Production Handoff	• Hardware Configuration Index released • Fusefile released	Complete the HPA Production Handoff Checklist	Word, Excel, Visio	• Completed Production Handoff Checklist	Production Handoff Checklist HC2 controlled	Production started

percentage should be defined. For example, process assurance may audit 15 percent of the peer reviews of test cases and procedures. When sample percentages are used, they should be earned—that is lowered or reduced only after a track record of execution to project plans and standards has been demonstrated. Process assurance should also define oversight performed when design or verification activities are performed by external organizations or subcontractors.

The process assurance plan should also describe how deviations to project plans and standards are recorded and tracked. Any forms used to record and resolve deviations should be referenced. Authorizations and approvals for deviations and their closure should be included in the HPAP.

HARDWARE STANDARDS

Hardware standards are the mechanism by which a company, through a formalized process, establishes the tangible characteristics of its products such as quality, reliability, durability, and consistency. Hardware standards can be used to foster a culture that promotes high quality and creativity. While the use of standards may not be required per DO-254 for DAL C and DAL D designs, the existence of hardware standards can indicate that a company has demonstrated a desire and/or commitment to standardize its design environment in a way that will engender consistent high quality in its products. Hardware standards are also used to assess the acceptability and quality of hardware design results by providing specific criteria to use in evaluations.

FAA Order 8110.105 clarifies the use of standards for HDL-based designs. The Order states that complex electronic hardware use coding (design) standards for HDL-based designs. The Order also specifies a review of the hardware requirements standards, design standards, validation and verification standards, and archive standards in the Hardware Planning Review, or SOI #1.

If standards are invoked for a project, then the standards become part of the certification basis and plans for the project. This means that the standards are part of the official certification data. Once invoked, a demonstration of compliance with the standards is needed. It behooves applicants and developers to have realistic and cost-effective standards. Training for the standards within the development group and to any sub-tier suppliers may be needed to ensure compliance by everyone concerned.

Tools can be used to enforce standards. Examples would include:

- Design rule checker in schematic editor/printed circuit board netlister.
- Fan-in, fan-out constraints during PLD synthesis/place and route.
- Code coverage tool to enforce verification completion standards.

Conversely, standards can be used to limit the scope of tool qualification. If a simulation tool requires qualification, the scope would be limited to the HDL functions and constructs permitted by the HDL design standards.

The standards should be referenced or invoked by the plan that uses them. The Hardware Design Plan should reference the requirements and design standards. The Hardware Validation Plan and Hardware Verification Plan should reference the validation and verification standards. The Hardware Configuration Management Plan should reference the hardware archive standards.

Well-crafted requirements and design standards can also be used as the basis for the validation and verification standards. The requirements standards will contain the criteria for creating, tracing, and justifying derived requirements. These criteria can then be used in the requirements validation review checklist. The requirements standards will contain the criteria for creating and tracing requirements. The criteria for the requirements can then be used in the requirements verification review checklist. The design standards will contain the criteria for creating the design and tracing it to the requirements. The design criteria can be used in the design verification review checklist.

Requirements standards include instructions for all aspects of writing requirements, validating requirements, using requirements management tools, tracing and validating derived requirements, and how to feed derived requirements back to the system design and safety process. The requirements standards should explain how to write requirements and include aspects such as the language and terminology, sentence construction, and definitions and applications of keywords such as "shall," "will," and "should." The requirements standards should also explain requirements feedback and clarification:

- Identifying requirements to feed back
- Where to feed back the requirements
- Methods for feeding back requirements
- Assessment criteria
- Artifacts to produce

The methods for requirements capture should also be explained. This can include top-down functional decomposition, rapid prototyping with phases to capture requirements, and spiral or cyclic development where requirements are developed and captured with more detail each time through the cycle. The standards should also describe the organization and layout of the requirements document.

Hardware design standards describe the procedures, rules, and methods for the conceptual and detailed design. The design standards also describe the guidance and criteria to design the hardware. The standards should include instructions for the design representation, i.e., standards for how the design is captured. Design standards should include:

- HDL
  - Allowable HDL types (VHDL, Verilog, ABEL)
  - Type/revision of HDL libraries and packages
  - Architectural considerations such as hierarchical or flat
  - Coding style
  - Coding standards
- Schematic
  - Schematic capture software and tool version
  - Library types

    - Schematic structure
    - Page size
    - Symbol design, look, and feel
- Other
    - Supporting representation such as flow diagrams, algorithm expression, state transition diagrams
    - Pen and ink drawings

Design standards may also include naming conventions for signals at all levels of the design. Use of intellectual property (IP) cores should be described in the standards. The design standards may also describe how to assess design alternatives and perform trade studies. Trade studies considerations should include:

- When to conduct a trade study
- How to conduct a trade study
- Criteria for the trade study
- Relative priority for trade study criteria
- How to assess results
- How to select alternatives

The design standards can also include guidance to assess fail-safe and fault-tolerant design features. The guidance should specify:

- Characteristics of fail-safe and fault tolerant designs
- When to assess the design features
- How to assess the design features
- Assessment criteria
- Evaluation methods/criteria
- Selection criteria

Design tool guidance should be described in the design standards. Topics to cover include which tools to use, how to configure the tool, constraints to be used, directives to include, reduction algorithms (if selectable), and place/route constraints for PLD designs.

Component selection criteria can be discussed in the design standards. Criteria can include technology options such as transistor–transistor logic (TTL), low-voltage TTL (LVTTL), complementary metal oxide semiconductor (CMOS), or low-voltage CMOS (LVCMOS). Proper selection ensures compatibility with other circuits and can minimize the complexity of the power supply design. Device temperature range is largely a function of packaging and device selection option. Care should be taken to respect maximum rated transistor junction temperatures. Guidance can also discuss how to select PLD types—CPLD versus FPGA, FPGA versus ASIC. The component selection can also describe how to select appropriate programming technology such as flash, anti-fuse, and static random access memory (SRAM).

VALIDATION AND VERIFICATION STANDARDS

Validation and verification standards include guidance, criteria, and methods for the validation of derived requirements and the verification of the requirements and its design. The validation of derived requirements may be covered in the requirements standards, the validation standards, or some combination of the two. As suggested above, the criteria in the requirements and design standards can be used as the basis for the review criteria used in the validation and verification procedures. A convenient method for performing reviews is to list the applicable criteria in a checklist or spreadsheet, then list the requirements, design, or verification artifact they are applied to.

Verification standards describe the criteria for the verification activities including review, analysis, and test. Reviews are applied to all life cycle data such as plans, standards, requirements, design data, analysis data, and test data. Reviews for life cycle data should check that the content described in Section 10 of DO-254 is documented. For example, a PHAC should, at a minimum, contain the information described in Section 10.1.1 of DO-254, a Hardware Design Plan should contain the information described in Section 10.1.2 of DO-254, and so on. The review criteria should be described or referenced in the verification standards. Review criteria are typically listed on a checklist and applied to the data or document under review. The checklist can also provide a method to document the review and any associated action and closure. Review procedures should be described in the verification standards. The procedures describe who can perform a review, how to perform the reviews, and when the reviews should be performed.

The verification standards also provide the guidance and criteria for verification by analysis. Criteria for circuit and timing analysis should be included in the standards. This can include component derating criteria to assess function and performance with component tolerance, voltage, and temperature effects. When simulation is used, the standards should explain test case selection criteria—the combination of inputs and initial conditions needed to comprehensively verify a requirement. Test case selection should include normal or expected inputs as well as abnormal or unexpected inputs. Tests would also include factors such as glitches on inputs, asynchronous resets, and timing variations on clock domains—especially when two or more clock domains are involved. The standards should also describe the expected coverage achieved with the elemental analysis. Note that test coverage for standard components will be different than coverage of a complex device such as an ASIC or an FPGA.

Finally, the verification standards also provide the guidance and criteria for verification by test. The standards should explain test case selection criteria—the combination of inputs and initial conditions needed to comprehensively verify a requirement with in-circuit electrical test. Test case selection should include normal or expected inputs as well as abnormal or unexpected inputs. Tests would also include factors such as glitches on inputs, asynchronous resets, and timing variations on clock domains—especially when two or more clock domains are involved. Note that the ability to achieve incorrect, abnormal, or unexpected inputs with hardware testing is typically limited due to the physical constraints of the actual circuit card.

HARDWARE ARCHIVE STANDARDS

The archive standards work in conjunction with the hardware configuration management plan. The standards are used to establish and then maintain an archive of life cycle data. The standards should address the type of archive used, such as optical disc or magnetic tape. Considerations such as medium selection and refresh periods should be specified. Archives should include data from all tools and databases used in the hardware life cycle process, including data used in the manufacturing process. The archives should also include any servers used to store data or drawings. When backups are used as part of the archive process, the backup content and frequency should be specified. Archive integrity checks should also be specified. Integrity checks could include message-digest algorithm, such as MD5.

If an organization does not have published standards, a well-designed peer review checklist can serve the same function. For this approach to work, the review checklist should contain detailed and comprehensive review criteria that express the same standards that would otherwise be found in a standards document. The checklist can also serve as guidance for the development of its target review item. For example, a design review checklist would contain review criteria that would capture all of the organization's design standards. Designers could use the checklist as a design standard by ensuring that the design meets all of the review criteria, and then the verification team could use the checklist as the basis for their design review where the design is confirmed to meet all the review criteria/standards. This approach provides a double layer of assurance that the design will meet its governing design standards.

SUBMITTALS AND COORDINATION

Table A-1 of DO-254 designates the Plan for Hardware Aspects of Certification, Hardware Verification Plan, Top-Level Drawing, and Hardware Accomplishment Summary as the documents submitted directly to the FAA. In some cases, the HVP is not submitted if an adequate description of verification is included in the PHAC and FAA Aircraft Certification Office (ACO) specialists are given a project familiarization presentation or meeting.

A top-level drawing is used in DO-254 Table A-1 since DO-254 is written for all electronic hardware in the aircraft equipment such as a line replaceable unit (LRU). In the context of how DO-254 is written and was conceived, the top-level drawing makes sense. Since compliance to DO-254 is only required for programmable electronic hardware in Advisory Circular 20-152, the FAA requests that a hardware configuration index (HCI) is submitted in lieu of the top-level drawing. FAA Order 8110.105 Section 4-5.a describes the use of the HCI document.

Once the planning is complete, it is recommended that all team members read and become familiar with the contents of the plans and standards. Company training can also be conducted with slide presentations to help facilitate the familiarization process and learning curve. These meetings are also ideal opportunities for management to express their backing of the plans, standards, and processes. The plan reading and training can also extend to subcontractors or service provider organizations.

In short, planning is not a short process. Four of the DO-254 objectives are for planning. Many hours of work can be optimized or used in the most effective manner with adequate planning. Planning, and especially the Plan for Hardware Aspects of Certification, is project specific. While design, verification, process assurance, configuration management, and standards may be common to projects within an organization, it is always beneficial to take time at the beginning of a project to assess them. Effective planning can also incorporate lessons learned from prior programs and fine-tune the processes.

REFERENCES

1. Order 8110.49 CHG 1, *Software Approval Guidelines*, Federal Aviation Administration, dated September 28, 2011.
2. IEC TS 62239, *Process management for avionics—Preparation of an electronic components management plan*, Geneva, Switzerland: International Electrotechnical Commission, 2003.
3. Order 8110.105 CHG 1, *Simple and Complex Hardware Approval Guidance*, Federal Aviation Administration, dated September 23, 2008.

4 Requirements

Learning to write requirements is similar to learning a new language: it takes time to learn a language, but with practice progress will be made. This learning experience applies to both individuals and organizations. Organizations may also have to experience a shift in culture as they adopt effective requirements writing and reading skills. Many practices and habits may need to change to adopt more effective and useful styles of writing and organizing requirements.

The information presented in this section on requirements is the culmination and synthesis of years of experience working on requirements writing and management. Best practices have been learned and gathered from numerous aerospace projects that span organizations working across the globe. After many years and numerous discussions, it became apparent that some ways of capturing, conveying, and organizing requirements were more effective than other methods. It also turns out that using this way of capturing, conveying, and organizing requirements streamlines the verification process and increases the test coverage of the design. Traceability became easier, readability improved, and comprehension of system functions was more apparent.

Traceability efforts and data can also be more streamlined and useful when the structure of the requirements mirrors the structure of the design, or in other words, if the requirements are written and decomposed from top to bottom and are organized by function, their structure will naturally default to a form that will match the functional structure of the resulting hardware. Ideally, requirements are isomorphic with respect to the design they describe.

Most important, there is a potential for more effective and more efficient verification when the requirements, the starting point for verification, is approached with the requirements tailored to that purpose. While it may seem counterintuitive, requirements that are written to be verifiable are always appropriate for designing, but the reverse is not always the case. The one exception to this rule is requirements that are written according to the requirement concepts presented in this chapter, since those requirements will be written to be equally appropriate for both design and verification. Thus in the general sense it is arguably more effective for requirement authors to get into the habit of writing (or learn to write) requirements with verification, rather than designing, in mind.

The authors do not intend to convey that there is a right or wrong way to capture or specify requirements. Rather, the idea is to convey a paradigm that is compatible with the objectives and guidance of ARP4754, DO-254, and DO-178. The ideas expressed pertain mostly to functional requirements, although it would be possible to apply the ideas to other types of requirements as well.

The paradigm presented here is often difficult for engineers and designers. Engineers and designers like to solve problems and design solutions. The educational system for engineers strongly emphasizes design, while requirements and functionality are often overlooked. Thus training may be needed to update people's background and reference experiences with requirements. Companies can organize internal training that introduces new or updated requirements standards and company templates for requirements. Training is also an effective way to get a team to work in a consistent manner, to work through differing perspectives, and begin to apply new skills.

To work well with DO-254, requirements should specify what the hardware must do and not how it should do it. This is the fundamental reason for requirements: to capture the intended functionality of the hardware and provide the starting point for both the design and verification activities. However, documenting functionality is often a less valued behavior or skill set for an engineer. Adopting a black box perspective on a system or part and describing what it does rather than describing the details of how it does it can be a challenge. As it turns out, a functional perspective can also be a learned or acquired skill. It also turns out that designers do think about the intended function when they design a circuit, but the "what" is often subservient to the "how."

Appreciating the difference between "what" and "how" can be confusing at first, particularly if a person was trained in an engineering culture that stressed the "how" over the "what." However, there are analysis techniques that can facilitate the discrimination between functionality and implementation, and at the same time facilitate the understanding of the underlying concepts.

The way in which requirements are used, as well as the way in which they are written, is often influenced—or even dictated—by the prevailing engineering culture. Requirement writing skills are often learned from examples in the workplace: engineers use requirements documents from previous programs as the starting point, and may also look at customer or industry documents. While examining other requirements documents is a helpful practice, the issue of whether the requirements are well formed is often not considered. In the absence of any other guidance on requirements capture, the engineer will often assume that the requirements are well written or that their requirements should be written in a similar, or identical, manner.

Requirements are central to the design and verification processes. The design is developed from the requirements, traces to the requirements, and is peer reviewed against the requirements. The verification is requirements based—test cases for hardware testing and simulations are based on the requirements. Traceability is established between requirements, test cases, test procedures, and test results. These relationships are depicted in **Figure 4.1**. Using a well-formed set of requirements as the entry into design and verification will ensure that the resulting design complies with the requirements and that the verification demonstrates that the requirements were met. Rework of requirements due to ill-formed starting conditions can result in redesign and/or rework of the verification.

Ideally, requirements capture should start before the design begins. The hardware design is generated from the requirements. A design based on requirements helps ensure that the intended function is implemented. Extreme care must be used if a

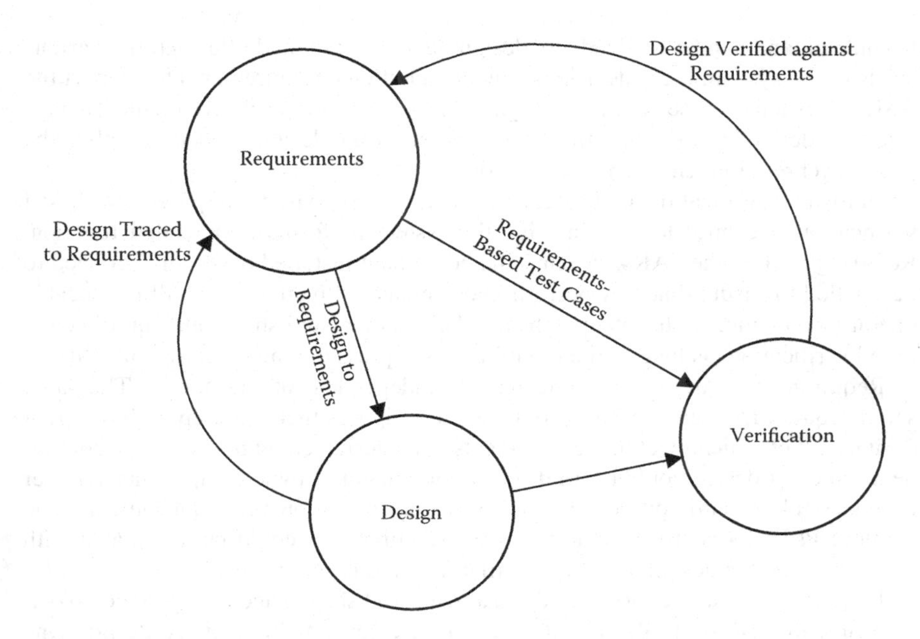

FIGURE 4.1 Central Role of Requirements

design exists and the requirements are backfilled or reverse engineered; this scenario often creates requirements that describe the existing design rather than the intended function.

WHY USE REQUIREMENTS?

The certification process and the regulations applicable to system and equipment installed on an aircraft state that the system must perform its intended function when installed on the aircraft. The industry has adopted the notion of requirements to express the aircraft functions. Thus, the use of requirements is needed to demonstrate compliance to the aviation regulations.

Requirements also inform the designer about what they need to design. Many designs are possible, but they may or may not be the correct design or a design that meets the requirements. Also, because verification relies on requirements-based testing, the use of requirements that describe the design rather than the design's function will result in design-based testing, which will only prove that the design is the design. It will not prove that the design performs its intended function, which is the ultimate goal of verification as defined in DO-254.

Requirements allow functions to be established and described from the highest levels. The requirements are allocated down through the various levels until they are realized as a design. For example, ARINC outputs from a system are described at the system level, then at the circuit card level where they are generated, then at the software and PLD level where they are designed and implemented. The functions and their requirements flow all the way down to the level at which the design is formulated and can be realized. As design decisions are made through the various levels of the system, these decisions can add additional requirements to support the design decisions. These additional requirements are known as derived requirements. They are derived in the sense that they flow out of the design needed to realize the upper level requirement(s) and its function.

Hardware verification in DO-254 is requirements based. Requirements-based verification is central to showing that the system performs its intended function. Referring back to the FARs, systems and equipment installed on an aircraft need to be verified to ensure that they perform their intended function under all foreseeable operating conditions. The intent of the regulation is accomplished with requirements-based verification (as long as the requirements express the intended functionality).

Requirements can also be conceived as independent of the design. The same set of requirements can be given to different designers that come up with different designs, all of which meet the requirements. One design could be all analog components, another design could be all discrete combinatorial logic components, another design could use a microprocessor and software, while a fourth design could use one or more PLDs. Any design that meets the requirements could be acceptable with the caveat that the design meets the availability and failure probabilities.

Requirements can be formulated, organized, and shaped according to the associated function. DO-254 talks about functional elements and associated design. One very effective method of organizing a set of requirements is to group them by the functions or functional elements. In other words, put all the requirements for a function in one part of the document. The requirements within that group can be further organized to express the behavior of all the outputs from the function. Skillfully writing and organizing requirements can produce a set of requirements that is structurally similar to the design, or isomorphic, and facilitates the verification. The technique of organizing requirements by functional element, then the outputs within the functional element, uses the architectural principle of letting form follow function. Writing requirements with a specific sentence structure also leads to requirements that are isomorphic with respect to the design and verification processes.

Finally, requirements describe what a system or function does, not how it works. Understanding the functionality allows a good fit with the hazard assessment and safety processes. Separating the requirements from the design also allows the mind to develop distinct internal models about what a system or function does that is separate from the design or how the system does what it does. One technique that works well in keeping these abstractions distinct and separate is to imagine the system or function as a geometric object where only the inputs and outputs can be observed and described. The art of writing requirements is to keep the attention and description limited to the inputs and outputs of the function. This will also guarantee that the function can be verified through review, analysis, and test.

REQUIREMENTS AUTHOR

Too often, requirements specifications contain design or implementation requirements, which can make verification very difficult or even ineffective. Thus the author of the requirements should ideally be able to write about the functionality without describing design or implementation details. Normally this will be the person who knows the most about what the hardware item needs to do, which in turn is often the person who is responsible for the next higher level of hardware (parent hardware) because that is the person who determines how the hardware item will function and what its role will be. The design decisions at the parent hardware level will define virtually all of the requirements in the child hardware, especially the requirements that define how the item will interface and interact with the parent hardware, and that define how the item will contribute to the parent hardware's functionality.

PLD requirements would ideally be written by the circuit card designer who selected the PLD because the card designer normally will know most about what the PLD must do in the circuit card. The best time to capture the requirements for a PLD is while the circuit board design is in work because the circuit card designer is working out the details of the PLD's role in the card while looking at the data sheets for all the circuits that the PLD is connected to. The circuit card designer also does not yet know how the PLD will be designed, so the PLD requirements will tend to be more functionally oriented.

Similarly, the requirements for the circuit board are ideally written by the assembly or system designer. As with the PLD, the designer of the upper level will know what functions are to be performed by the circuit card and the signals that the circuit card connects to.

For requirements that describe function rather than design, the requirements author does not need to be a competent circuit card design engineer to write the requirements for the circuit card. Although understanding circuit card designs may help the requirements author and ensure that the requirements are realistic and realizable, it can also bias the requirements toward design and implementation. Similarly for a PLD, the requirements author does not need to be a competent PLD design engineer to write the requirements for the PLD. On the other hand, requirements that competently describe a design must be written by a competent designer.

Figure 4.2 shows the generalization of this concept. The requirements are ideally captured by the person that works on the design of the next higher level of abstraction within the system. Note too that not all systems will have or need all the levels shown in Figure 4.2. CCA in the figure stands for circuit card assembly.

In general the item's designer is not always the best person to write the item's requirements. The item's designer is likely to know the most about the technology to be designed, but as previously stated, such competence is not a prerequisite for writing good functional requirements, and in fact may bias the requirements toward implementation over function. In addition, the item's designer may not be the person who knows most about the item's intended functionality. There is certainly no official guidance that discourages an item's designer from writing his or her own requirements, but experience has shown that it is not always the best path to high quality requirements because it often results in incomplete requirements (requirements written by the designer may lack detail because the designer is too close to the topic and will not need the details for his or her own use of the requirements) as well as requirements that capture intended design rather than intended functionality (designers often look ahead and specify what they intend to design rather than look back at what the design has to do, plus they may not have an extensive understanding

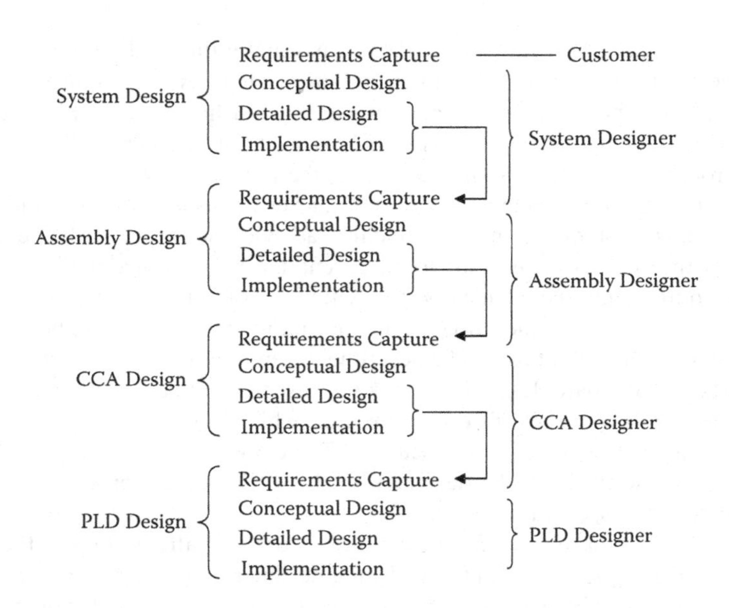

FIGURE 4.2 Requirements Capture

of how the item will integrate into the parent hardware). Overall, it is just a better idea to have the person who knows most about what the item has to do write the requirements.

However, as with any rule, there are always exceptions. In this instance, the designer may be an acceptable requirements author if they also designed the parent hardware or collaborated with the parent hardware's designer to either design the parent hardware or to define the functionality that gets allocated to the item, so they will fully understand the item's intended functionality. Without this understanding the designer may have a tendency to fill in the gaps in their functional knowledge with intended implementation.

SYSTEM REQUIREMENTS

System requirements development is governed by the processes and activities described in ARP4754 and are the starting point for DO-254 processes. Requirements for systems are driven by many factors. The system safety processes will generate performance constraints for the availability and integrity of a function. The availability expresses the continuity of a function while the integrity expresses the correctness of a function. The availability and integrity will become drivers for various architectural features of a design. These features will then generate requirements. It is helpful (and necessary) to tag or specifically identify these safety specific requirements to allow them to be traced down through the design to the actual implementation.

System processes will also gather the intended function for a system as requirements that specify the desired system performance under the associated operating conditions. The functions for a system are created from a combination of customer wants, operational constraints, regulatory restrictions, and implementation realities. Customer requirements will vary with the type of aircraft, the function of the aircraft (cargo or passenger, business jet or commercial transport), and the type of system. Customers may also drive requirements based on the intended payload of the aircraft, knowledge of routes that will be flown, operating practices used by the customer or airline, maintenance practices used by the customer or airline, and any desired features. Operational requirements are used to define the functions for the flight crew, maintenance crew, and support staff. The operational requirements will consist of the actions, decisions, information, and timing for all personnel interactions and interfaces.

Performance requirements define the attributes of a system that make the system useful to the customer and aircraft. Performance requirements will define expected system performance, system accuracy, fidelity, range, resolution, speed, and response times. Physical requirements specify the physical attributes of the system and include size, mounting provisions, power, cooling, environmental restrictions, visibility, access, adjustment, handling, storage, and production constraints.

Maintenance aspects can also add requirements for systems. Scheduled and unscheduled maintenance may have different types of requirements. Some maintenance may be certification driven in order to meet system safety failure probabilities. Maintenance requirements should specify actions and interfaces used to access and test the equipment.

Interfaces for power and signals also add requirements to systems. System designs need to be compatible with allocated power sources and associated power budgets. Interfaces can also have specific impedance and electrical characteristics that need to be met for proper operation. Interface characteristics are requirements and they are different than the definition of signaling through an interface. The meaning of data or signals on an interface is captured in an interface specification. The interface specification is meant to be a definition and should not be treated as requirements. Instead, the requirements should specify how all signals designated as system outputs in the interface specification are generated and how all signals designated as system inputs in the interface specification are used.

Last but not least, the certification process itself may add specific requirements for a system. The certification regulations or certification authorities may specify features, attributes, or specific implementations. These types of requirements are typically added to show compliance with airworthiness regulations.

Once the system requirements are gathered, they can be sorted into various types: physical, environmental, performance, functional, and so on. The process starts with the requirements that originate at the aircraft and system level. These requirements are validated through review, analysis, and even testing to determine whether they are the correct requirements and whether they are complete. The initial validation of all the requirements is performed in the scope of ARP4754. The entry into DO-254 assumes a validated (complete and correct) set of system requirements. The entry into DO-254 also assumes that the initial concept for the allocation of functions to software, electrical components, mechanical components, and electronic components such as PLDs has been done.

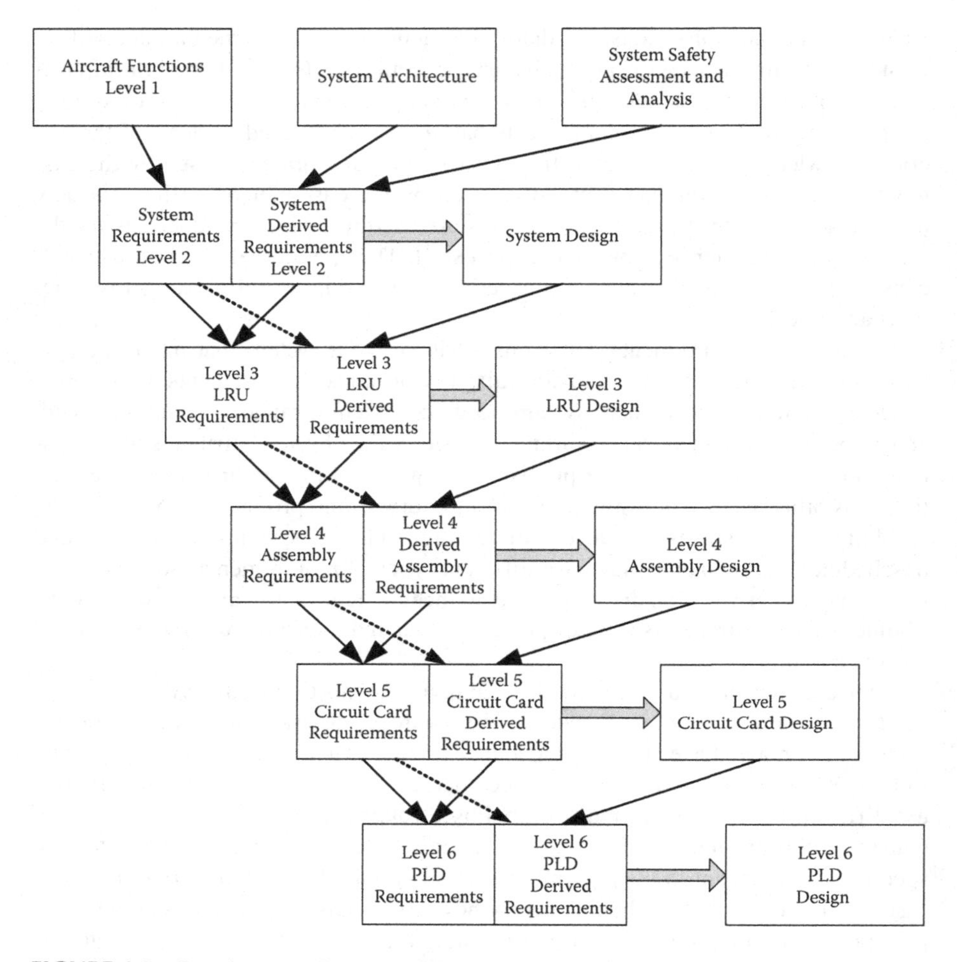

FIGURE 4.3 Requirements Decomposition

As the system design evolves, additional requirements are added to support the design decisions made along the way. For example, a system level decision that chooses to implement some of the system functions in software may add hardware requirements for a microprocessor and how it interfaces with peripheral circuitry. The software development process itself will add requirements as the software architecture, design, and packaging is chosen. Systems that need redundancy and/or dissimilarity will introduce additional requirements due to the architecture and topology.

Figure 4.3 shows the decomposition of functional requirements and the additional derived requirements added during the design process.

TYPES OF REQUIREMENTS FOR ELECTRONIC HARDWARE

Within the requirements allocated to the electronic hardware, there are some categories or types of requirements worth noting. These categories were developed based on observations of what engineers were actually putting in their requirements documents. Repeated occurrences of general types of requirements led to the categorization to help requirements authors understand what types of requirements are most effective and what types are not. These categories include:

- Functional requirements
- Implementation requirements
- Application requirements
- Input requirements
- Indirect requirements
- Referred requirements
- Global requirements
- Derived requirements

Functional requirements are the most recommended type of requirement. Functional requirements describe behavior and can be expressed as how an output responds to an input (or time event). In addition, functional requirements are testable, expressed in quantifiable terms, and are independent of the design and implementation of the hardware.

Implementation requirements describe specific aspects of the hardware design or implementation. Implementation requirements are not recommended as they skip the functionality and instead describe how to design or implement a function. Requirements that describe a specific design or implementation are not testable; they usually have to be verified through an inspection. Implementation requirements also compromise the integrity of the design and verification processes because there is simply no way to verify intended function. The verification will only prove that the design is the design or that the implementation is the implementation. If specific design or implementation features are important, they can be captured as textual information in the requirements document and not treated as a formal requirement. Some design or implementation features can also be elaborated in hardware design standards which also separates them from the requirements. The design or implementation features can also be documented in the hardware design documentation. Putting these details in the hardware design documentation allows them to be used to drive lower level derived requirements, if required.

Application requirements are used by requirements authors to describe how the hardware will be used. While this is good information, it should not be expressed as requirements. Information about the application of the hardware is beyond the scope of the functionality of the hardware. In addition, the application cannot be verified within the scope of verifying the functionality. While information about how the hardware is used is important, it should be documented as textual description and not as a requirement. The application should be considered, but any specific functions required for the application should be captured as functional requirements for the hardware.

Requirements for inputs are also often written. As with other nonfunctional requirements, input requirements are not recommended. No amount of design associated with an input requirement can cause the input to happen. Requirements should express how the hardware responds to the presence or occurrence of an input. Input behavior is important and can be captured as descriptive text or in an interface specification. However, a requirement for the electrical characteristics of an input port (such as input impedance) is different than an input requirement. The input characteristics can be expressed as a requirement and can be verified through test and other means.

Indirect requirements imply functionality rather than stating it directly. Indirect requirements often use keywords such as "capability," "mechanism," or "provision." The problem with an indirect requirement is that it can be satisfied without actually implementing the function. Indirect requirements are also too ambiguous for a rigorous and methodical design and verification process. Indirect requirements can readily be converted to functional requirements by simply removing the "capability," "mechanism," or "provision."

Requirements authors sometimes try to take shortcuts or increase efficiency by referring to other documents for requirements. These referred requirements can create problems for traceability and maintenance, especially if the referenced document changes. In addition, the referenced document may not meet the same control category for configuration management as the original document. Sometimes authors even reference textbooks for requirements; this is very problematic if the textbook goes out of print.

Global requirements are often used in documents. These are readily identified since they describe behavior from an input to a group of outputs, such as how a reset input will affect all of the outputs. While sentences with global requirements are easy to write and easy to understand, they do not necessarily make effective requirements. Global requirements will force the reader to go through an entire document searching for every signal that affects an output or function of interest. The functionality must then be synthesized from separate data points spread throughout the document. Global requirements can cause requirements conflicts, perpetuate errors and mistakes, and make it difficult to understand intended functionality.

Finally, derived requirements are another type of requirement, but they differ from the previously discussed requirement types because they do not have any clearly defined characteristics that are unique to the statements themselves, or in other words, requirements cannot be identified as derived just by looking at the statements. Derived requirements can only be identified when they are compared to higher level requirements. "Derived" is a label that is attached to requirements of all types to identify them as requirements that need to be validated—any of the previously described requirement types can be labeled as derived, and that designation does not change the requirement in any way.

Derived requirements are requirements that result from the design process. They normally originate from design decisions that occur during the decomposition of system functions to lower levels of electronic hardware, and can be identified most easily by looking at their content in comparison to the higher level requirements: if the information in the lower level requirement contains specifications that are new or different, or if they specify new functionality, then they are derived.

Derived requirements are singled out by DO-254 for special treatment, and deserve lengthy discussion because they can be problematic if not addressed properly. Derived requirements are closely related to the topic of Validation, so they are discussed in detail in the Validation chapter.

The most important distinction between the above requirement types is the difference between functional and implementation requirements. Functional and implementation requirements are the two requirement types that are arguably the most and least compatible with the goals of DO-254 processes.

When trying to understand the difference between functional requirements, which specify what the hardware should do, and implementation requirements, which specify how the hardware should be designed, it can be helpful to look first at how requirements fit into the engineering design process, and at the same time expand on the roles that function and implementation play in the design process, how they are most often confused, how they should be documented, and how to get from implementation to function.

As defined by DO-254, the four major phases of the standard design process are requirements capture, conceptual design, detailed design, and implementation, with these four phases intended to be executed in chronological order. Astute readers will immediately recognize that requirements capture occurs at the start of the process. This should be a clue as to how requirements are supposed to be used: if the requirements are supposed to be written before the design is started, then it follows that requirements should only define functionality ("what") rather than design implementation ("how") because specifying the design requires jumping ahead to the conceptual detailed design phases.

Probing further into the relationship between requirements and the design process, since the "how" of the hardware cannot be known until after the design has been conceptualized or designed, and the requirements are constrained to contain only the "what" of the hardware's functionality, then the way to document the "how" of the hardware is to use another type of document that is written specifically for that purpose. That document is the Design Description document, sometimes referred to as a Design Document.

The design document captures the "how" of the hardware by documenting the hardware's detailed design. It describes how the hardware was designed or how it should be implemented, and has the following features:

- It is written after the hardware has been designed.
- It documents the details of how the hardware works.
- It documents the design implementation.
- It documents how the hardware meets its intended functionality.
- It does not contain requirements.

In contrast to the design document, the requirements document:

- Is written before the design is started.
- Specifies the hardware's intended functionality (what the hardware is supposed to do).
- Does not specify how the hardware should meet its intended functionality.
- Does not specify how the hardware should be designed.
- Does not document the design.

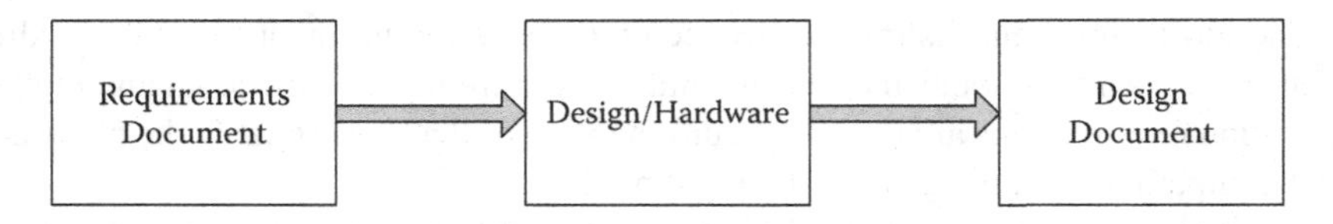

FIGURE 4.4 Requirements and Design Documents

The requirements document and the design document—and by extension, functionality and implementation—are essentially opposites. Both are important, although it can be argued that the design document is optional while the requirements document is essential, especially considering the processes in DO-254. Since the two documents are so different and serve such different purposes, then it makes sense that they not be combined or mixed together.

The difference between the two document types is straightforward, but again, depending on the engineering culture, the difference can be confusing at first. Using the design process again as a guide, if the design document is used to document the design, then it would appear near the end of the design process, in contrast to the requirements document, which appears at the beginning. If seen on a timeline, the two document types should bracket the design phases of the process as shown in **Figure 4.4**.

Ironically, in the real world it is a fairly common practice to combine the two documents into a hybrid document that is then misnamed a requirements document. There are numerous reasons, most of them well-intentioned, but most are due to not fully appreciating the difference between function and implementation and how they can affect the processes in DO-254, nor the difference between a requirement, a design description, and a design decision.

The difference between requirements, design descriptions, and design decisions can also be confusing at first, again depending somewhat on past training. The differences can be summarized as follows:

- A requirement defines a specific function that the hardware must implement.
- A design (implementation) decision is what designers do to get from the requirements to the end hardware design.
- A design description describes the hardware design, and should not contain specifications.

A very simple example of a card level functional requirement would be:

```
"DAT_OUT(11:0) shall output the 12-bit unsigned binary
equivalent of the AIN_1 analog input."
```

The above statement specifies what the output should do, not how it should do it. It contains no design information that will lead the designer to a specific solution, and it is written in the pre-design tense.

The design decision that results from this requirement might be to use a commercial 12-bit analog to digital converter integrated circuit (IC) to convert the analog input into a 12-bit unsigned binary value. It could also be to create a custom analog to digital converter out of transistors and passive components, or use a commercial 16-bit analog to digital converter and use only the upper 12 bits. Any of these solutions will meet the intended functionality, and because the requirement is written to express the intended functionality rather than its implementation, any of them (and more) can be considered as solutions.

Design decisions do not have to (in fact, they should not) be documented as requirements. Design decisions can be important, but it is the intended functionality of the design that is most important and most appropriate to DO-254 processes. Design decisions and the circuits that come from them should be documented in a design document.

The design description describes what the designer created to implement the function described in the requirement. It describes what was designed to allow the hardware to meet the requirement. It is written in the post-design tense and does not contain specifications. For the analog to digital converter example, the design description might contain the following:

> "The `AIN _1` input signal is buffered, scaled, and level shifted by an OP123 operational amplifier that converts the +/–10 Volt input to the 0–5 Volts required by the ADC123 16-bit analog to digital converter. The ADC123 is configured to generate an unsigned binary output by strapping the `MODE(1,0)` input pins to DGND. The upper 12 bits of the ADC123 digital output are passed through an LD123 tristate line driver to the `DAT _OUT(11:0)` output pins."

The design decision for this design might be:

> "The ADC123 converter was selected because it exceeds the performance needs of this function, and because it is in widespread use throughout the electronics industry it can be purchased at considerably less cost than the ADC234 12-bit converter IC. The OP123 is used in this buffering role because it was designed to work with the ADC123 and likewise exceeds the performance needs for this function. The alternative of using the ADC321 converter IC was considered since its input range will accommodate the full voltage range of the `AIN _1` input signal, but its cost was considerably greater than the cost of using the ADC123 and OP123 combination, and there was some concern as to whether its performance over temperature would be adequate."

Using the design description as a requirement is a common but unadvised practice. For example, "There shall be a 12 bit analog to digital converter configured for an unsigned binary output" might be a typical implementation requirement for the analog to digital converter example. While this approach to requirements can provide tight control (more likely too tight) over the design of the hardware, it does not capture the intended functionality of the hardware. In fact, such a requirement has no intended functionality at all because it only describes how the hardware should be designed. Alternatively, the design description will be almost identical to the requirement because both the requirement and the design description are describing the same thing. Thus a simple test to determine whether a requirement is functional or implementation is to compare the requirement to a description of the design. If they are similar or identical it means the requirement is capturing implementation rather than intended functionality.

The intended functionality in an implementation requirement is not often readily perceived, even by the person who wrote the requirement. In fact, it is often very difficult to determine the intended functionality even when the specified circuit is analyzed. However, one method to get from the implementation to the correct intended functionality is to perform a "root function" analysis. A root function analysis is similar to the well-known root cause analysis, and has a similar procedure: repeatedly ask, "why?" to identify the root functionality that is driving the specification.

The following example illustrates how an implementation requirement can be converted into its root functionality (and converted into a functional requirement) through this process.

Starting with the following card level requirement: "The ADC shall sample the `AIN_1` analog output at least 20 microseconds after and 20 microseconds before a transition on any motor drive output."

This requirement is clearly an implementation requirement—it tells the designer that there is an analog to digital convertor (ADC) on the card and that it should sample its input at a certain time. Underlying all implementation requirements is functionality that the implementation is meant to fulfill, and that functionality is what the requirement should be specifying. In this case there is a reason for causing the ADC to sample its input at the specified times, but the reason and the functionality that it serves is not clear from the requirement itself.

To get to the reason and its intended functionality the requirement can be analyzed through the root function process by asking a series of "why" questions. The first "why" question would be, "Why does the ADC have to sample its output at least 20 microseconds before and after a transition on a motor drive output?" The answer to this question would be, "The motor drive outputs are high current and can cause significant distortion in small signal analog circuits, which will cause conversion errors in the ADC. Sampling at least 20 microseconds from the motor drive output transitions will allow the ADC to convert the `AIN_1` input after noise has settled and allow time for the ADC to convert the input before the next transition."

The "why" portion of the answer, or in other words the part of the answer that actually provides the justification for the requirement, is the second sentence—that the ADC needs to convert its input after the motor drive transition noise has settled and with enough time to complete a conversion before the next transition.

If the original requirement is then rewritten to reflect the justification in the second sentence, the new requirement would state, "The ADC shall sample `AIN_1` after motor drive transition induced noise has settled, and complete its conversion before the next incidence of motor drive transition induced noise."

This requirement is a little more explanatory than the original in that it focuses more on the reason for sampling 20 microseconds before and after a motor drive transition rather than blindly specifying that it do so. However, while it is closer to a functional requirement, it is still more of an implementation requirement because it focuses on how the circuit should work rather than the functionality that it should satisfy.

If another "why" question is asked, i.e., "Why should the ADC sample the input after the motor drive noise has settled and with enough time to convert before there is more noise?" The next answer is "The noise caused by the motor drive outputs will cause errors and distortions in the ADC's digital output. Avoiding the noise will minimize the errors and distortion in the ADC output."

So if the requirement is rewritten again from the justification in the previous answer, it could be, "The ADC shall sample the `AIN_1` input in a way that minimizes errors and distortion in its digitized output." This requirement is better yet, but it still does not capture the root or core functionality that it is really trying to express.

So another question is in order: "Why does the ADC need to minimize errors and distortion?" The answer to this question is, "If the conversion errors in the ADC output are not minimized, the total error in the digitized output may exceed 2%, which is undesirable."

If the requirement is again updated to include the last answer, it could be written as, "The total error in the digitized `AIN_1` output shall be less than 2%."

Note that at this point the requirement is free of implementation: there is no mention of circuit architecture, components, or anything else that hints at a physical implementation. It very simply specifies an input/output/performance relationship that defines the functionality of the output (the "what" of the output) without delving into how it should be done.

An additional check is to ask one more "why" question about the final requirement: "Why must the digitized `AIN_1` output have less than 2% error?" The answer to this question is, "If the `AIN_1` digitized output has more than 2% error the software system control loop that uses the value will become too inaccurate and potentially unstable."

This last answer is significant in that its reason or justification now points to a higher level than the card, so the justification—and any requirement written from it—is now outside the scope of the hardware item (the card). This difference from the earlier answers tells us that the justification for the card level requirement has now gone high enough to be free of the implementation of the requirement. This is the sign that the requirement now captures the intended functionality at the card level, and that this intended functionality is flowing down from the parent hardware rather than being generated at the card level.

The process for determining underlying functionality in an implementation requirement is displayed in tabular form in **Table 4.1**.

The final form of the requirement specifies a limit for errors in the output and does not restrict the sources of the errors even though the original requirement focused on errors due to high current transients. This is a significant improvement in the effectiveness of the requirement, and one of the reasons that implementation requirements are not as effective as functional requirements: the system's justification for the requirement was that the system needed the error in the digitized output to be kept below 2%, and without regard to the source of those errors. In contrast, the original requirement focused only on one potential source of errors out of the many different sources that exist in every electronic system, and some of those other sources could be more problematic than the one being addressed.

TABLE 4.1
Finding the Root Functionality

Requirement	Question	Justification	Justification Level
The ADC shall sample the AIN_1 analog output at least 20 microseconds after and 20 microseconds before a transition on any motor drive output.	Why does the ADC have to sample its output at least 20 microseconds before and after a transition on a motor drive output?	Sampling at least 20 microseconds from the motor drive output transitions will allow the ADC to convert the AIN_1 input after noise has settled and allow time for the ADC to convert the input before the next transition.	Card
The ADC shall sample AIN_1 after motor drive induced noise has settled, and complete its conversion before the next incidence of motor drive induced noise.	Why should the ADC sample the input after the motor drive noise has settled and with enough time to convert before there is more noise?	Avoiding the noise will minimize the errors and distortion in the ADC output.	Card
The ADC shall sample the AIN_1 input in a way that minimizes errors and distortion in its digitized output.	Why does the ADC need to minimize errors and distortion?	If the conversion errors in the ADC output are not minimized, the total error in the digitized output may exceed 2%.	Card
The total error in the digitized AIN_1 output shall be less than 2%.	Why must the digitized AIN_1 output have less than 2% error?	If the AIN_1 digitized output has more than 2% error the software system control loop that uses the value will become too inaccurate and potentially unstable.	System

If the ADC circuit was designed as specified by the original requirement, it would only be designed to address one of those many sources, and in the end the method specified in the requirement might well have been inadequate to actually keep the error below 2%. The final design could meet its requirements and pass all of the requirements-based verification tests and still not meet its intended functionality, or in other words its performance would not meet the needs of the system.

Requirements that capture root functionality have significant advantages over requirements written to capture implementation and design features: (1) the requirements will express functionality rather than implementation, which will ensure that the hardware item meets the needs of the system; (2) they will express the functionality at the most appropriate level of detail for the hardware item; (3) they will express the functionality that is required or intended by the next higher level of the system in a form that is compatible with those needs, and thus be more true to the needs of the system; (4) the functionality in the requirements will have a clear connection (or trace) to the higher level functionality from which it flowed, which will enhance both validation of the requirement and an understanding of how the functionality of the item relates or integrates into the functionality of the system; and (5) the requirements will not restrict how the functionality can be implemented, allowing the real experts—in this case the card level designer—to perform their job to the best of their ability. In addition, requirements like these will express functionality in a way that can be measured and verified.

Another consideration in how requirements should be written is whether the contents of a requirement are actually essential (i.e., required) for the hardware to meet its goals (the intended functionality as defined or flowed down from the system level).

A requirement specifies a quality or parameter without which the hardware item cannot correctly or completely fulfill its role in the system. DO-254 alludes to this concept when describing validation activities, in which it states that requirements are complete when all of the attributes that have been defined (i.e., specified) are necessary, and all necessary attributes have been defined. Thus if a specification is not necessary, then it probably should not be a requirement.

In contrast to requirements, definitions contain information that is non-essential but still helpful for understanding the hardware item's function, and for designing or verifying the hardware, such as descriptions of input signals, definitions of terminology, definitions of modes of operation, or a preferred method of implementation.

There can be a tendency in requirements authors to write too many requirements, or in other words, to specify more than they need to specify. One contributing factor is that they may confuse or blend requirements and definitions and capture definitions as requirements. This practice is understandable, but ultimately it can cause needless work by creating more requirements than are necessary and specifying more than needs to be specified.

One common example of crossing the line between definitions and requirements is the concept of modes of operation. Operating modes are often captured in requirements because humans are used to conceptualizing hardware operation in terms of modes. There is no question that using modes of operation is useful and even essential in some arenas, but despite their importance they are actually definitions that are good to define but are not always appropriate as requirements.

Consider the following requirement that defines a mode based on the states of four input signals:

```
"The Control box shall be in Spoiler Mode when the LOC(3:0)
input signals are set to 1010 binary."
```

This requirement seems useful enough by defining a mode of operation so the designer will know what the box is supposed to be doing when the inputs are in the defined states, but it really should not be a requirement: conditions that are dictated by the states of inputs are definitions, and a requirement like this one cannot be verified because there is nothing that can be measured to prove that the box is in Spoiler Mode. After all, the inputs can be measured to prove that they are in the 1010 states, but that does nothing to prove that the hardware is actually in a Spoiler Mode, whatever that may be.

If we compare this requirement to the definition of a requirement as stated previously, the information in the example requirement is non-essential, and the hardware will work just fine without declaring that the state of the `LOC(3:0)` inputs is called *Spoiler Mode*.

Modes that are specified as a condition of outputs are more appropriate because the outputs can be measured to verify that the hardware is in a particular mode, but even in this situation the definition of modes of operation is still not essential for the hardware to meet its goals. The reason is that modes are just an abstraction, a name given to a particular condition in the hardware that humans use to help them understand what is going on with their hardware. The hardware itself has no knowledge of modes of operation; it just responds to its inputs the way it is designed to. Humans, however, can more readily conceptualize and relate to the hardware's operation if modes are assigned to its various functional states.

While it may not be advisable to define modes with requirements, it can be advantageous to define modes (but not with requirements) and then use the modes in requirements. Using modes in this manner allows the requirements to exploit their advantages (segregating operation into easily comprehended states) without needlessly increasing the number of requirements that have to be managed.

For example, if there is a fairly complex functional state that is characterized as `(LOC(3:0) = 1010) AND (SER_MES(65) = 1) AND (OPSTATE(4:0) = 11001) AND (BIT_SEL(6:0) = 0010111)`, the requirements for the hardware can be greatly simplified and much easier to understand if a name is assigned to this functional state and then used in requirements in place of the actual signals. Thus instead of writing the following requirement:

```
"The ROT_UP output signal shall assert high when the EXT_UP
input signal is high and LOC(3:0) = 1010 AND SER_MES(65) = 1
AND OPSTATE(4:0) = 11001 AND BIT_SEL(6:0) = 0010111."
```

The requirement can be written as a definition combined with a requirement, as follows:

```
"The hardware will be defined as being in the EXTEND mode of
operation when (LOC(3:0) = 1010) AND (SER_MES(65) = 1) AND
(OPSTATE(4:0) = 11001) AND (BIT_SEL(6:0) = 0010111)."
"The ROT_UP output signal shall assert high when the EXT_UP
input signal is high and the hardware is in the EXTEND mode."
```

By adding this mode definition (and ostensibly defining a number of other modes as well) the operation of the hardware is more easily understood in terms that are conducive to human understanding as well as being in accord with the overlying functionality of the hardware.

So what is the issue with creating requirements for things that really do not need them? Cost is the main issue: each requirement creates a cascade of activities, each of which comes with a time, cost, and schedule impact. Every requirement has to be written, archived, reviewed, validated, traced, verified, and maintained throughout the life of the product. This cost can be significant, especially when the number of requirements is large. Reducing the number of requirements can have a significant impact on the long-term cost and schedule of the program.

Note, however, that reducing the number of requirements does not mean cramming more information into a fewer number of requirement statements. The cost of requirements is proportional to the number of attributes, not necessarily the number of statements. Writing huge requirements that each have 100 attributes will not cost less than writing 10 times as many requirements that each have 10 attributes.

ALLOCATION AND DECOMPOSITION

Requirements and the functions they express flow down from the aircraft level to the design that realizes the function. Aircraft functions become the requirements for the aircraft systems—the functions are divided up and allocated to various aircraft systems. Each aircraft system has a requirements specification. If the system contains numerous parts, there may be a requirements specification for the controller or line replaceable unit (LRU). When the controller or LRU is initially designed, decisions are made as to how to implement the functionality and the type of architecture in the electronics that would best suit the purpose. This is the initial step in allocating system functions to mechanical or electrical components, software, or electronics such as an FPGA or an ASIC.

Initial efforts in programs using PLDs used system requirements that flowed directly to PLD requirements. For an extremely simple system this could be possible. With complex systems and highly integrated functionality, an allocation from system directly to an FPGA creates problems with traceability and verification. A more thorough approach would be to start with system requirements and allocate through the various levels of abstraction and design within the system. One such approach would start with the aircraft functions and system architecture. The next lower level of abstraction would be system requirements and associated design. The following level would be LRU (controller) requirements and associated design. The next level would be assembly requirements and associated design. The next level would be circuit card requirements and associated design. Finally, there are PLD requirements and the PLD design. The flow down of requirements is depicted in **Figure 4.5**.

The idea is that the functions get allocated or decomposed from a higher level of requirements to a lower level of requirements. A function could be the transmission of ARINC data. The ARINC outputs are visible at the system level and described in the system level requirements. The design and allocation of the ARINC transmission function does not happen until the circuit card level requirements are reached. At the circuit card level, the requirements can then be decomposed and allocated to the electronics and even software if needed.

Requirement flow down is performed by allocating the requirement from one level down to the next level. If the function is the same at both levels, the requirement can simply be copied. If the terms in the requirement change units or split into

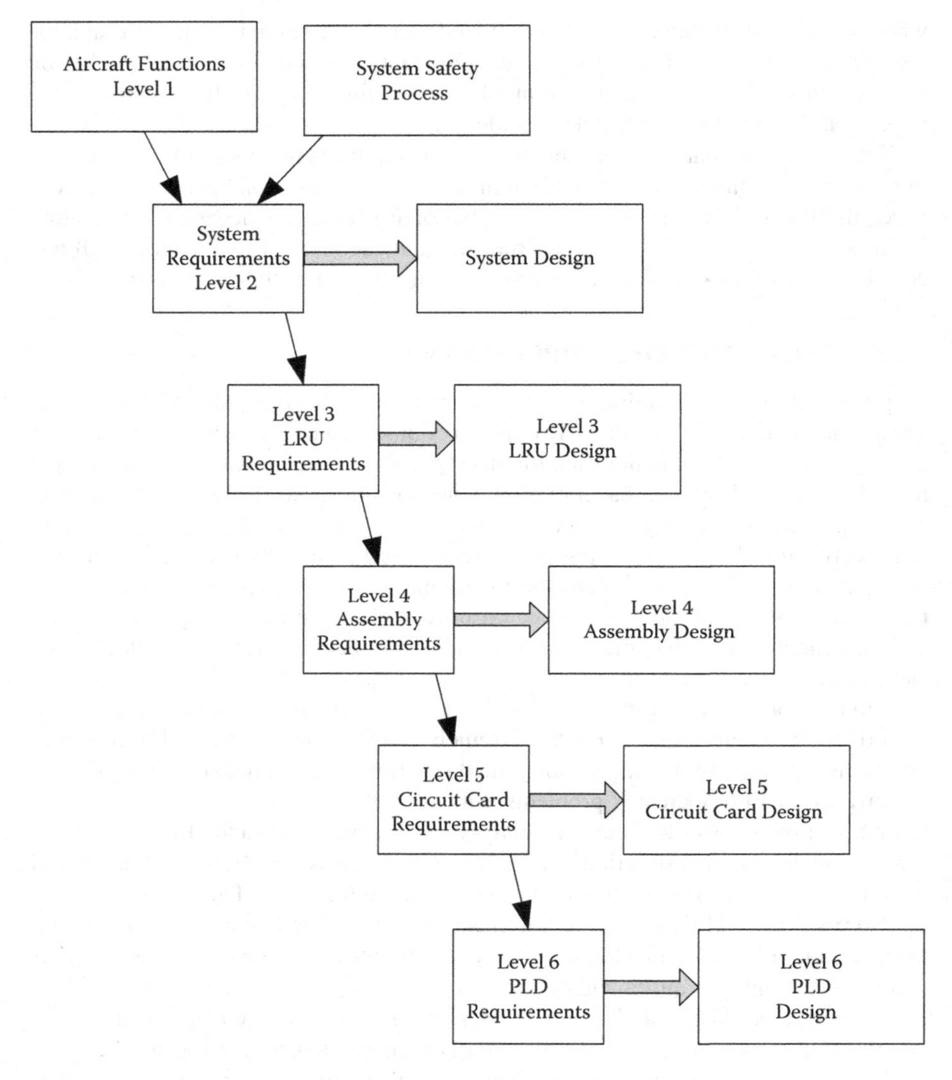

FIGURE 4.5 Requirements Flow Down

parts, then the requirement decomposes into more granular parts. For instance, a level n requirement can be the same on two or more levels. The requirements would look like this:

- System level: The ARINC 429 output bus shall transmit at 100 Kbits-per-second.
- Circuit Card level: The ARINC 429 output bus shall transmit at 100 Kbits-per-second.

If the terms in the requirement change units, then the requirement decomposes and takes on the relevant units. A requirement can be similar on two levels with different units. The requirements would look like this:

- System level: The maximum flap output position command shall be limited to 30.0^{0}.
- Circuit Card level: The maximum flap output position command shall be +15.0 V +/– 0.03 V.

The requirements can decompose between levels, with parts allocated to software, electronic components, and PLDs. In this case the requirement will change between levels both in terminology and in the number of requirements. This example shows an upper level requirement decomposing into five lower level requirements:

System level:	The LRU shall transmit Flap Position with Label 204 at a 50.0 Hz +/– 0.1Hertz rate.
LRU level:	The `ARINC 429` outputs shall transmit Flap Position with Label 204 at a 50.0 Hz +/– 0.1 Hertz rate.
Circuit Card level:	The `ARINC 429` outputs shall output Flap Position with Label 204 at a 50.0 Hz +/– 0.1 Hertz rate.
PLD level:	The `ARINC_429_N` and `ARINC_429_P` outputs shall generate a message with Label 204 when a write operation is performed to address 0x30334024 on the processor data bus. The message output shall start within 1 microsecond after the completion of the write operation, and use the data from the write operation as the message payload.
Software level:	`Label_204_FLAP_POSITION` shall be written to `ADDR(0x30334024)` every 20 +/– 0.04 milliseconds.

In this example, at the circuit card level the functionality in the requirement is essentially unchanged from the LRU level because in this fictional system the ARINC messages are generated on the circuit card before being output unchanged through a connector on the LRU, so the output of the circuit card and the output of the LRU are the same. At the next lower level, the rate and payload functionality are allocated to the software while the functionality for generating the actual message and inserting the message label and data is allocated to the FPGA that generates the ARINC message. The FPGA has no knowledge of the system rate and data requirements; it only knows that when the processor writes a data word to a particular address that it should generate an ARINC message with a particular label and use the data from that write operation as the message payload. The software likewise has no knowledge of the particulars of the ARINC messages, or even of the existence of the ARINC transmitter function; it only knows that it has to write a particular data item to a particular bus address every 20 milliseconds. The combination of the functionality in the software and FPGA requirements satisfies the higher level requirements at the circuit card, LRU, and system levels.

The circuit card requirement is allocated rather than derived because its functionality flowed down unchanged from the LRU level. On the other hand, the FPGA and software requirements that were derived from design decisions made at the circuit card level (the decisions to generate the messages in the FPGA and generate the data and timing in software) are derived requirements that trace to the circuit card requirement and therefore need to be validated. Or more accurately, the two FPGA requirements need to be validated but the software requirement must be addressed as specified in DO-178.

Functions (and their requirements) can also originate at any level of the hierarchy, although this is often indicative of incomplete requirements capture in the upper levels. These requirements are added to allow testability or other features desirable for electronics or software development, but are not necessarily a system function. Suppose that the software developers want to add a test feature. They could add a requirement at their top level of requirements. An example might be:

- `Level 5 (Software): The software shall output the status of monitor1, monitor2, and monitor3 every 100 milliseconds on ARINC maintenance label 350.`

TIMING AND PERFORMANCE SPECIFICATIONS

It can be tempting for a requirements author to base timing and performance specifications on the capabilities of the technology or hardware rather than on a flow down of system performance requirements. For example, the author of a PLD requirement might specify the input to output timing of a signal from an understanding of how fast the PLD can generate that output rather than from the needs of the parent circuit card, or in cases where the requirements are being written after the design is underway, from a knowledge of how long the signal actually takes. From the perspective of how the PLD actually performs this requirement may seem more accurate, but because it captures only the actual performance of the hardware rather than the performance that the system needs from the PLD, it is in effect disconnected from the system and its performance. Relying on such methods can eventually result in performance specifications that are both inaccurate from the system perspective and problematic from the practical perspective.

Like other aspects of functionality, timing and performance specifications should flow down from higher level requirements and design decisions, with the exception of derived requirements that originate at the item's level and are not related to system functions. The specifications should not be influenced by the characteristics of the hardware.

For example, if a PLD has to output a clock that is used to drive an analog to digital converter (ADC) IC on the parent circuit card, the specifications for the clock output should be derived from the clock specifications in the ADC data sheet combined with any system requirements for timing and performance. The speed of the PLD, the frequency of its input clock from the card, and the type of PLD should have no bearing on the timing and performance specifications because those parameters should flow down from higher levels to ensure that they are determined by system needs.

When specifications are flowed down from higher levels they will normally be expressed as a value with a tolerance or as a range of acceptable values, and those values and tolerances are dictated by what the system needs. For example, an FPGA requirement for a clock output to an ADC device should be determined by a combination of how many conversions must be performed and the performance limits of the ADC. In turn, the number of conversions that must be performed will flow down from the system as a number of analog inputs and how often each of those inputs needs to be converted, which will then define how much time can be allotted to each conversion, which will then define how fast each conversion must be performed, which will then define a range of clock frequencies that will satisfy the conversion performance while remaining within the capabilities of the ADC. The FPGA clock frequency does not factor into the derivation of the requirements because that is an implementation detail at the card level that should be determined by the requirements for the ADC rather than the other way around.

Decomposition of a requirement and its trace to a lower level requirement can follow several possibilities. The requirement can be copied directly, one upper level requirement to one lower level requirement. An upper level requirement can break out into two or more lower level requirements. Two or more upper level requirements can trace down to one lower level requirement. Two or more upper level requirements can trace down to two or more lower level requirements, although requirements decomposition and traceability work better when the 1:1 or 1:many types of relationships are used. In summary, requirements decompose as follows:

- 1 upper level requirement : 1 lower level requirement
- 1 upper level requirement : 2 or more lower level requirements
- 2 or more upper level requirements : 1 lower level requirement
- 2 or more upper level requirements : 2 or more lower level requirements

The decomposition relationships are shown in **Figure 4.6**.

One enhancement of the requirements process is to capture the traceability data when the requirements are captured. This is more effective and efficient than waiting and trying to figure it out at a later time. Trace relationships should be defined by the flow down and decomposition of functionality (requirements) from higher to lower levels. Since traceability is the natural byproduct of the hierarchical design process, the best time to document the traceability is when the requirements are being written so that the traceability and functional flow down will be faithfully captured. Writing requirements and then generating traceability later on as a separate activity is not recommended.

WRITING REQUIREMENTS

Requirements are used to identify and specify the functionality of an aircraft, a system, an LRU, a circuit card, a programmable logic device, and software. There is more to a requirement than the word **shall**. A sentence with the word **shall** in it is not necessarily a requirement let alone a well-written requirement.

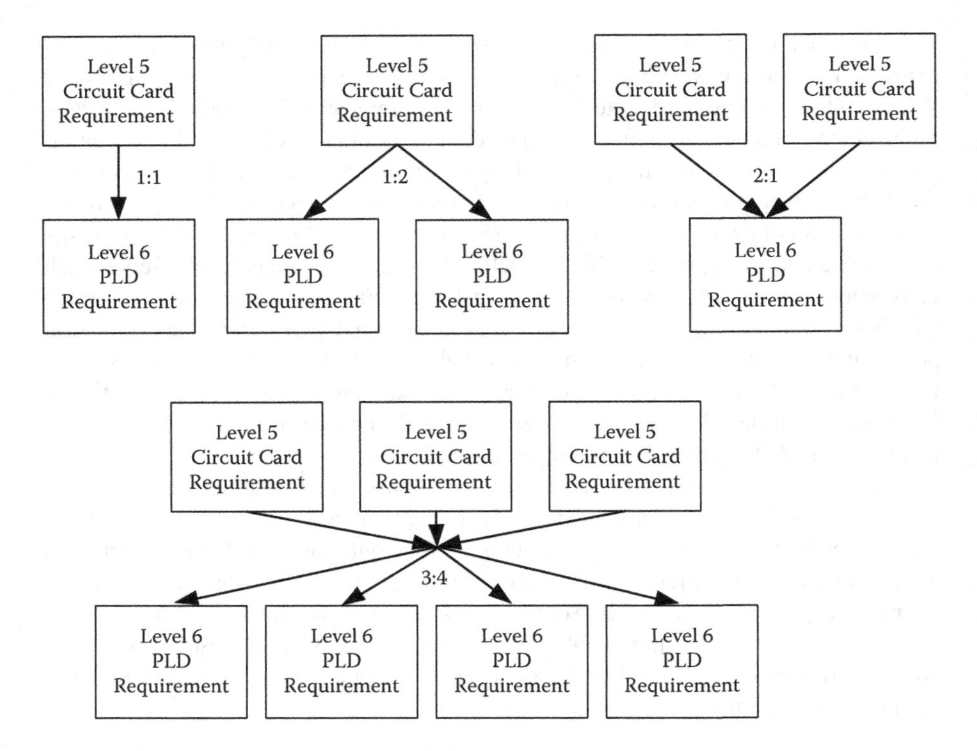

FIGURE 4.6 Requirements Decomposition

Requirements should be expressed in the positive sense. Or, facetiously, the requirements document shall not use negative requirements. Negative requirements are difficult to verify; it is hard to prove that something does not happen.

Requirements also express a cause and effect relationship between the output(s) and the input(s) or the output(s) and a timing event. The timing event can be expressed as absolute timing, such as 20 microseconds, or relative timing such as 20 microseconds before the next rising edge of the clock.

The following are recommendations to help write well-formed requirements:

- Use the keyword "shall" to identify a requirement
- Make each requirement unique
- Assign a unique identifier or tag to each requirement
- Specify what is done, not how it is done
- Specify an observable behavior on an output in the requirement
- Describe how the outputs respond to valid inputs
- Describe how the outputs respond to invalid inputs
- Describe how the outputs respond to timing related events
- Make each requirement specific and concise
- Express requirements in the positive sense, no negative requirements or "shall not"

- For PLDs or circuits, express requirements in terms of pin level behavior
- Use descriptions of outputs that can be observed and inputs that can be controlled

Write the requirements for the target audience—the design engineer and the verification engineer. Requirements should not be written to show off the literary prowess of the author. The difficult part is for the requirements author to not use their assumptions, biases, or presuppositions when writing the requirements. Instead, the author should aim to impart their knowledge to the reader and explain the details. Writing for the target audience is enhanced with the following techniques:

- Define behavior from **output to inputs**, not input to outputs.
- Explicitly specify the behavior of the output(s) rather than implying the behavior.
- Put all relevant information in one place for each output or function. Do not force the designer to search through the entire document to get the information. Do not use global requirements where one input affects all the outputs.
- Define **functionality**, not implementation. The requirements should focus on what the hardware does, not how it does it.
- Use information statements to explain the application or other contextual information. Information statements can also contain definitions for specific terminology and keywords. If certain aspects of the design or implementation are important, then describe it in an information statement.

The author should add all details necessary to explain the functionality. Make no assumptions about what the reader already knows or about what assumptions the reader will make. Fill in the details in the requirements rather than forcing the reader to add details or figure things out. Use a glossary or designated section of the document to define terminology. Use the defined terminology in a consistent manner throughout the requirements document.

Requirements should define timing dependencies when they apply. Use tolerances for specific timing, such as "10.0 microseconds +/– 0.1 microsecond." Another method to specify timing is to use a limit such as "within 5.0 microseconds" or "before the next rising edge of the input clock `MSTR_CLK`". Specify timing with tolerances for all aspects of behavior. Provide measurable reference points for all timing specifications.

Requirements need to tell the reader the order of precedence so that behavior can be predicted in the presence of simultaneous inputs or events. Using language that conveys precedence lets the reader know which behavior occurs when a reset occurs and whether the reset overrides other inputs. Denoting a word such as "unconditionally" as the highest precedence will allow the requirements to specify order of importance. Another way to indicate precedence is to define the sequence in which inputs or events are processed or responded to. The sequence can be in a list or defined in a table.

While requirements standards are part of DO-254, they also enhance requirements capture and interpretation. Using a standard for requirements will result in a standardized document format. This allows team members to readily know where to find information in a requirements document. Team members can more readily contribute to a project or get added to the team when the requirements documents have a standard content and format. The standards also help new employees understand how to make technical contributions: they do not have to try to come up with documentation structure and presentation format because it is already defined. The standards define the meaning and usage of modal operators of necessity and modal operators of possibility, such as "should," "shall," "can," "will," "may," and "might."

The use of a template is recommended for requirements capture. The template helps the requirements author think through all aspects of functionality. If certain aspects of the template do not pertain to a function, then that part of the template can be marked not applicable. Templates also ensure consistency in how requirements are written and help enforce the requirements standards. Other benefits of using a template across an organization can be realized when engineers can readily contribute to a project without having to learn new requirements methods for new projects. A typical template could be expressed as follows:

Function Name
Definition of terminology
Trace links
Justification for derived requirements
Output(s)
- Description
- Units
- Encoding (bus, number of bits and weighting of least significant digit)

Input(s) affecting the output
Power-on behavior—the behavior of the signal when power is first applied
Reset response—the state of the signal after reset asserts or reset deasserts
Assert behavior—the requirements for asserting the output to its active state
Deassert behavior—the requirements for deasserting the output from its active state
Invalid behavior—how the signal should behave when presented with invalid or undefined inputs

Finally, the requirements should express a cause/effect relationship between inputs and outputs. The inputs (or timing event) are the cause, the outputs are the effect. This can be expressed diagrammatically as follows:

```
The {output or verifiable aspect}
  shall
    {effect}
      when {cause}
```

Once requirements are structured in this fashion, they support the design and verification process. Well-formed requirements may take additional time and effort, but it is time and effort well spent. Once a complete and well-formed set of requirements is created and reviewed, it is much faster and efficient to create a design to implement the functionality expressed in the requirements. The verification process for testing and analysis can proceed with the requirements before the design even starts.

PLD REQUIREMENTS

Requirements for PLDs can be structured to facilitate the design and verification processes. Since the inputs and outputs are digital with a value of 0 or 1 (false or true), the requirements expressions can be patterned to state under which conditions an output is true (1) or false (0). Another approach is to define when an output is asserted or deasserted. The requirements document will then define whether an asserted or deasserted output means high or low, 1 or 0, true or false. Writing requirements with outputs as a function of input follows the same structure of hardware design language for a PLD or a schematic for a circuit. Hardware design language uses statements to express how output signals are assigned when a sequence of events has occurred, when a timing event occurs, or when a reset occurs.

Using a requirements template as previously mentioned:

Function Name
- Output(s) of the function
- Input(s) affecting the output
- Definition of terminology
- Trace links
- Justification for derived requirements
- Power-on behavior—the behavior of the signal when power is first applied
  - Requirement(s) describing behavior of the output during power-on conditions
- Reset response—the state of the signal after reset releases
  - Requirement(s) describing behavior of the output during or immediately following reset conditions
- Assert behavior—the requirements for asserting the output to its active state
  - Requirement(s) describing behavior of the output(s) that are asserted when the input conditions are satisfied
- Deassert behavior—the requirements for deasserting the output from its active state
  - Requirement(s) describing behavior of the output(s) that are deasserted when the input conditions are satisfied
- Invalid behavior—how the signal should behave when presented with invalid or undefined inputs
  - Requirement(s) describing behavior of the output(s) when the input conditions are invalid or undefined

The structure of the template mirrors the structure of the HDL. The first requirement is for the response to a reset input. The first section of the HDL process decodes the clock and reset inputs and responds to a reset before processing normal inputs.

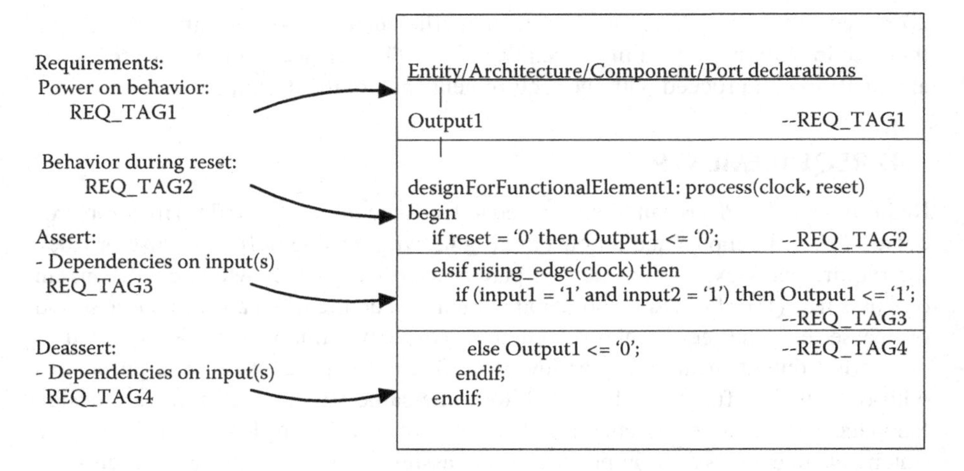

FIGURE 4.7 Structure of Requirements and Design

This also establishes the precedence of the reset over the other conditions shaping the output signal. The second requirement is to assert the outputs to their active state in response to the input conditions being satisfied. The second section of the HDL process decodes the clock and inputs and asserts the output if the gating conditions are satisfied. The third requirement is to deassert the outputs to their inactive state in response to the input conditions being satisfied. The third section of the HDL process decodes the clock and inputs and deasserts the output if the gating conditions are satisfied. When specific conditions to deassert the output are not needed, the deassert requirement could simply state to deassert the output when the conditions to assert are not satisfied.

Figure 4.7 shows the requirements structure and how it is set up to mirror the design.

PLD requirements can alternatively be stated with a similar template.

```
The {output or verifiable aspect}
  shall
    {effect}
      when {cause}
```

This can be expanded into a more complete set of aspects of a typical requirement:

```
The {output or verifiable aspect}
  shall
    {precedence modifier}
      {assert, deassert, set to value}
        {timing modifier}
          {absolute or relative time event}
            when {input condition or absolute time event}
```

Filling in the modifiers shown in the brackets, the structure becomes:

```
The {output or verifiable aspect}
  shall
    {always, unconditionally, only}
      {assert, deassert, set to value}
        {before, after, when, during}
          {xx nsec, the next rising edge of a lock, read/write
              asserts low}
            when {inputs are set to a combination of hi/low,
                 a sequence of events has occurred or
                 a timed period elapses}
```

The following shows an example of a requirement that fills in the above structure to create a requirement:

```
The {d_out[15:0] pins}
  shall
    {unconditionally}
      {be set to 0x0000}
        {within}
          {50 nanoseconds}
            when {reset_n is asserted low}
```

Compacting this into a more readable sentence:

```
The d_out[15:0] pins shall unconditionally be set to 0x0000
within 50 nanoseconds when reset_n is asserted low.
```

This is another example of a requirement that fills in the above structure, leaving not applicable (n/a) for terms not needed in the requirement:

```
The {d_out[15:0] pins}
  shall
    {n/a}
      {be set to the current_position_value}
        {within}
          {50 nanoseconds}
            when–
              {read_en_n is asserted low and
              cs_n is asserted low and
              addr_in[16:0] = 0x12C2} during a processor read
```

Compacting this into a more readable sentence:

```
The d_out[15:0] pins shall be set to the current_position_
value within 50 nanoseconds when read_en_n is asserted low,
cs_n is asserted low, and addr_in[16:0] = 0x12C2 during a
processor read.
```

The preceding requirement also shows how to express access to an FPGA register. Rather than explicitly using registers in the requirements (which can be thought of as a design detail) the access to an address is described in terms of write and read transactions. For a register write, the requirements state the resultant observable effect of having data on the data bus for the address corresponding to what will be a register in the design while the control signals are configured to write data to the FPGA. For a register read, the requirements state what data will be on the data bus for the address corresponding to what will be a register in the design when the control signals are configured to read data from the FPGA.

Compound conditions can be formulated and expressed in the requirement. This is typically the case when an output has a dependency on several inputs or time events. The following examples show how to combine conditions to express the dependencies typically found in PLD logic. Also observe that these structures match the logic gate or expression that they express.

An output that asserts when a set of input conditions must be met all at the same time can be expressed:

```
out1 shall assert high within 50 nsec when the following
conditions are satisfied:

  • Condition1
  • Condition2
  • Condition3
```

This structure is a logical and (AND) of the input conditions.

An output that asserts when any one of a set of input conditions is met can be expressed:

```
out1 shall assert high within 50 nsec when one or more of the
following conditions are satisfied:

  • Condition1
  • Condition2
  • Condition3
```

This structure is a logical or (OR) of the input conditions.

An output that asserts when none of the input conditions are met can be expressed:

```
out1 shall assert high within 50 nsec when none of the follow-
ing conditions are satisfied:

  • Condition1
  • Condition2
  • Condition3
```

This structure is a logical not-or (NOR) of the input conditions.

An output that asserts when at least one of a set of input conditions are not met can be expressed:

```
out1 shall assert high within 50 nsec when at least one of the
following conditions are not satisfied:
```

- Condition1
- Condition2
- Condition3

This structure is a logical not-and (NAND) of the input conditions.

An output that asserts when a set of input conditions must be all met or all not met at the same time can be expressed:

```
out1 shall assert high within 50 nsec when either all or none
of the following conditions are satisfied:
```

- Condition1
- Condition2
- Condition3

This structure is a logical exclusive nor (XNOR) of the input conditions.

An output that asserts when only one of input conditions are met can be expressed:

```
out1 shall assert high within 50 nsec when only one of the
following conditions are satisfied:
```

- Condition1
- Condition2

This structure is a logical exclusive or (XOR) of the input conditions.

ELECTRONIC HARDWARE REQUIREMENTS

An output data bus from electronic hardware can be described using techniques similar to those used for PLD requirements. The requirements first describe the physical behavior of the signal, then the logical aggregation is described. The output as a function of input structure is used again. The following requirements are an example of how ARINC output data requirements could be written. An ARINC bit is a high followed by a null or a low followed by a null.

```
MON_DBUS_P is an ARINC output bus that will have three output
values, identified as the low output value, the null output
value, and the high output value.

    a. [ARINC-OUT-200] MON_DBUS_P shall be set to -10.0 Volts
       +/- 300 millivolts when a low value is output on the
       ARINC bus.
```

b. [ARINC-OUT-210] MON_DBUS_P shall be set to 0.0 Volts +/- 300 millivolts when a null value is output on the ARINC bus.
c. [ARINC-OUT-220] MON_DBUS_P shall be set to 10.0 Volts +/- 300 millivolts when a high value is output on the ARINC bus.

[ARINC-OUT-230] The MON_DBUS_P rise time from null to high and low to null output values shall be 1.5 microseconds +/- 0.5 microseconds when a bit is output on the ARINC bus.

[ARINC-OUT-240] The MON_DBUS_P fall time from null to low and high to null output values shall be 1.5 microseconds +/- 0.5 microseconds when a bit is output on the ARINC bus.

[ARINC-OUT-250] MON_DBUS_P shall unconditionally assert to its null output value within 2.0 microseconds after the RESET input signal asserts to logic low.

[ARINC-OUT-260] MON_DBUS_P shall unconditionally remain at its null output value while the RESET input is asserted to logic low.

[ARINC-OUT-270] Each bit of the serial data stream output on the ARINC data bus shall be 10.0 microseconds +/- 5% in duration.

a. [ARINC-OUT-280] MON_DBUS_P shall output either a high output value or a low output value during the first 5.0 microseconds +/- 5% of each bit of the serial data stream.
b. [ARINC-OUT-290] The time for each high or low shall include the time needed to transition from the null to high or the null to low output values.
c. [ARINC-OUT-290] MON_DBUS_P shall output a null value during the second 5.0 microseconds +/- 5% of each bit of the serial data stream.
d. [ARINC-OUT-295] The time for each null shall include the time needed to transition from the high to null or the low to null output values.

[ARINC-OUT-310] MON_DBUS_P shall assert to the high output value during the first 5.0 microseconds +/- 5% of the bit when the data content of the bit being generated is logic high.

[ARINC-OUT-320] MON_DBUS_P shall assert to the low output value during the first 5.0 microseconds +/- 5% of the bit when the data content of the bit being generated is logic low.

[ARINC-OUT-330] MON_DBUS_P shall default to the null output value when no data is output on the ARINC bus.

[ARINC-OUT-340] MON_DBUS_P shall output 32-bit serial data messages for each LABEL shown in Table 4.2.

TABLE 4.2
ARINC Data Bus Messages

SSM	SDI	LABEL (Octal)	DATA	RATE
00	00	050	Position Feedback1	30.0 milliseconds
00	00	051	Position Feedback2	30.0 milliseconds
00	00	055	Position Feedback3	30.0 milliseconds
00	00	057	Position Feedback4	30.0 milliseconds
00	00	113	Ram Position Feedback	5.0 milliseconds
00	00	240	Power Status	50.0 milliseconds

[ARINC-OUT-350] MON_DBUS_P shall output a null with a minimum duration of 40.0 microseconds +/- 2.5% between successive 32-bit serial data messages.

[ARINC-OUT-360] Each message shall be output on the ARINC bus once within the period indicated in the RATE column in **Table 4.2**.

Complex functionality can also be expressed using the output as a function of the input format. The output is described as a transfer function to allow time domain or frequency domain characteristics to be added. The next example builds on the preceding ARINC output for the data field of Label 050. The note below (e) conveys the intent of the author without using an implementation requirement or stating the design explicitly as a requirement.

[ARINC-OUT-380] Message 050 shall have the content shown in **Table 4.3**.

[ARINC-OUT-390] The 16-bit digitized value in bits 11 through 29 of Message 050 shall be the binary representation of the FEEDBACK input signal (Digitized WING1POSFDBK), where 0xFFFF indicates a positive full-scale value corresponding to an input voltage of +10.00 Volts, and 0x0000 indicates 0.00 Volts.

Note: A filter is applied to the input for noise reduction and signal shaping.

The Digitized WING1POSFDBK will have the following characteristics:

a. [ARINC-OUT-400] The magnitude of Digitized WING1POSFDBK shall decrease linearly from the FEEDBACK input value when the input frequency is less than 100.0 Hertz.

TABLE 4.3
ARINC Message 050

DATA STREAM BIT	BIT FUNCTION	CONTENTS
1	Label msb	0
2	Label	0
3	Label	1
4	Label	0
5	Label	1
6	Label	0
7	Label	0
8	Label lsb	0
9	SDI	0
10	SDI	0
11	Data	0
12	Data	0
13	Data	Digitized WING1POSFDBK bit 0 (lsb)
14	Data	Digitized WING1POSFDBK bit 1
15	Data	Digitized WING1POSFDBK bit 2
16	Data	Digitized WING1POSFDBK bit 3
17	Data	Digitized WING1POSFDBK bit 4
18	Data	Digitized WING1POSFDBK bit 5
19	Data	Digitized WING1POSFDBK bit 6
20	Data	Digitized WING1POSFDBK bit 7
21	Data	Digitized WING1POSFDBK bit 8
22	Data	Digitized WING1POSFDBK bit 9
23	Data	Digitized WING1POSFDBK bit 10
24	Data	Digitized WING1POSFDBK bit 11
25	Data	Digitized WING1POSFDBK bit 12
26	Data	Digitized WING1POSFDBK bit 13
27	Data	Digitized WING1POSFDBK bit 14
28	Data	Digitized WING1POSFDBK bit 15 (msb)
29	Data Sign	0
30	SSM	0
31	SSM	0
32	PARITY	Odd Parity

b. [ARINC-OUT-402] The magnitude of the Digitized WING1POSFDBK shall be 3 dB +/- 1.0% down from the amplitude of the FEEDBACK input at 100.0 Hertz.
c. [ARINC-OUT-404] The magnitude of Digitized WING1POSFDBK shall decrease by 12 dB +/- 1.0% from the amplitude of the FEEDBACK input over each octave (doubling) of the input frequency when the input frequency is greater than 100.0 Hertz.
d. [ARINC-OUT-406] The phase of the Digitized WING1POSFDBK shall be within 2.0 +/- 1.0% degrees of the phase of the FEEDBACK when the input frequency is between 0.0 and 500.0 Hertz.

Note: The intent is for the output to be filtered with a 2nd order low-pass Butterworth function with a 100.0 Hertz –3dB cutoff.

[ARINC-OUT-410] Each bit of the Digitized WING1POSFDBK shall have been digitized from the FEEDBACK input no more than 1.0 millisecond before being transmitted in the serial data stream on MON_DBUS_P.

[ARINC-OUT-420] Bit 32 of Message 050 shall be set to logic high if the sum of the logic highs in bits 1 through 31 is even.

[ARINC-OUT-430] Bit 32 of Message 050 shall be set to logic low if the sum of the logic highs in bits 1 through 31 is odd.

REQUIREMENTS ORGANIZATION

Requirement documents benefit from providing a definition of the signals connected to the hardware. One approach is to first describe the hardware inputs, then describe the hardware outputs, and finally the bidirectional signals connected to the hardware. The definitions list the proper name for each signal. The signal name is then used consistently throughout the rest of the document. No requirements or "shalls" are used in the definitions. A pinout or connection diagram can also be included.

The next section of the document includes a textual description of the functionality of the hardware. Block diagrams should be included to show functional interfaces and how signals are grouped.

The requirements are then organized by what DO-254 calls a functional element. The simplified explanation is to organize the requirements according to the function. Within each functional element, list the output, then the requirements for each output. Organizing requirements by functional element will facilitate the functional failure path analysis and the elemental analysis, especially when using DO-254 as written, to cover all electronics. **Figure 4.8** shows the organization of requirements functional element, then each output.

The document should be balanced where balanced means that there are requirements for each signal listed in the signal definition section as an output or the output side of a bidirectional signal. Each signal listed in the signal definition section as an input or the input side of a bidirectional signal should be consumed or utilized in the conditions associated with generating one or more outputs.

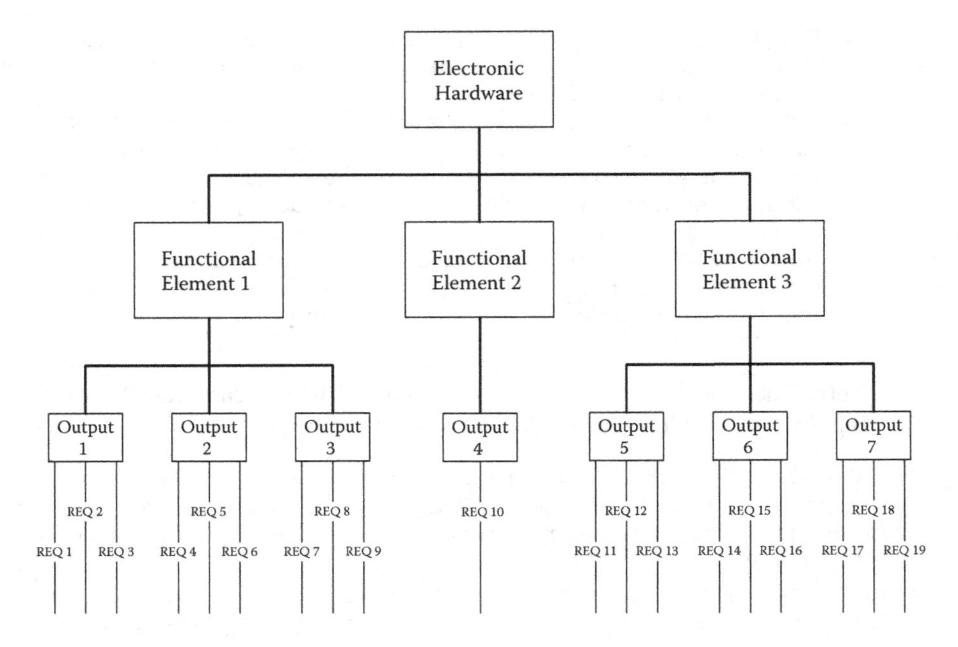

FIGURE 4.8 Requirements Organization

SYSTEMS, HARDWARE, AND SOFTWARE

While DO-254 was conceived and written for all complex electronic hardware, the current regulations only require compliance for programmable logic devices. Some programs and international certification authorities are starting to require compliance to DO-254 for all complex electronic hardware. One potential benefit of using DO-254 for electronics in systems is that it can pave a way through the dilemma of tracing software requirements directly to system requirements and PLD requirements directly to system requirements. Many times derived requirements are added to hold everything together and these derived requirements use the notion of derived meaning that there is no upward trace for a requirement. This style of derived requirements uses a paradigm of functionality being an orphan—not having a place to originate from.

A unified view of systems using ARP4754, DO-254, and DO-178 is now possible. The unified view includes the design decisions' connections to the associated derived requirements. "Derived" in the unified view takes on the original meaning as defined in the glossary of DO-178 and DO-254. The unified view of systems and requirements takes the following path:

- Aircraft functions are identified
- These aircraft functions are expressed as requirements
- The aircraft level requirements are allocated to systems
- The system requirements capture the system functionality and any additional requirements from the safety processes

- The system design captures the architecture and features necessary to implement and realize the system functionality, safety, and reliability
- Allocation and decomposition then continues down through one or more levels of abstraction
- Each level of abstraction copies (and repeats) its allocated requirements until there is a design
- The design will then show how requirements are decomposed
  - Divided up among software, electronics, PLDs
  - Creating a transform between different representational systems
    - Mechanical to electrical and vice versa
    - Electrical to electrical
    - Rotational to linear and vice versa
    - Three dimensions to two dimensions and vice versa
    - Logic bits or voltage into displays
  - Design decisions are captured and become the drivers for derived requirements at the next lower level of abstraction
- Software requirements are allocated and decomposed from the requirements for the circuit card that hosts the microprocessor or microcontroller
- PLD requirements are allocated and decomposed from the requirements for the circuit card that hosts the PLD device

Figure 4.9 shows this unified concept for requirements. With the software traced and associated with the electronic hardware that hosts it and through which it acquires inputs and sends outputs, there can be greater clarity and justification for software derived requirements. The software will be closely aligned with the hardware and software memory map, the input/output circuitry, and the necessary interfaces between the hardware and the software.

Using the same principles above for form following function, this unified approach has the software functions and requirements emerging from the design where it is realized. In other words, the software is part of the circuit card that hosts the microprocessor or microcontroller. Software functions are fulfilling circuit card or board level functions through the electronic hardware that the software interacts with. Software drivers can now trace to the hardware design decisions for components and circuitry. Software access to input/output read/write can also be tied to the memory map and hardware/software interfaces.

Most profoundly, this unified concept resolves a lot of the issues associated with the use of an inconsistent definition and usage of derived requirements as originally presented in Section 5 of DO-178B and the glossary of DO-178B.

The unified approach has system level outputs that are driven by LRU outputs. The LRU outputs are defined as the response(s) to the LRU inputs and timing events. LRU outputs are driven by circuit card outputs; the LRU inputs drive the circuit card inputs. Circuit card outputs are driven by the combination of software, PLDs, and other electronic circuits. The software, PLDs, and other electronic circuit inputs come from the circuit card inputs and any front end, processing, or conditioning circuits on the circuit card. With a consistent signal naming scheme, an alignment of these signals and their decomposition through various layers of abstractions is possible.

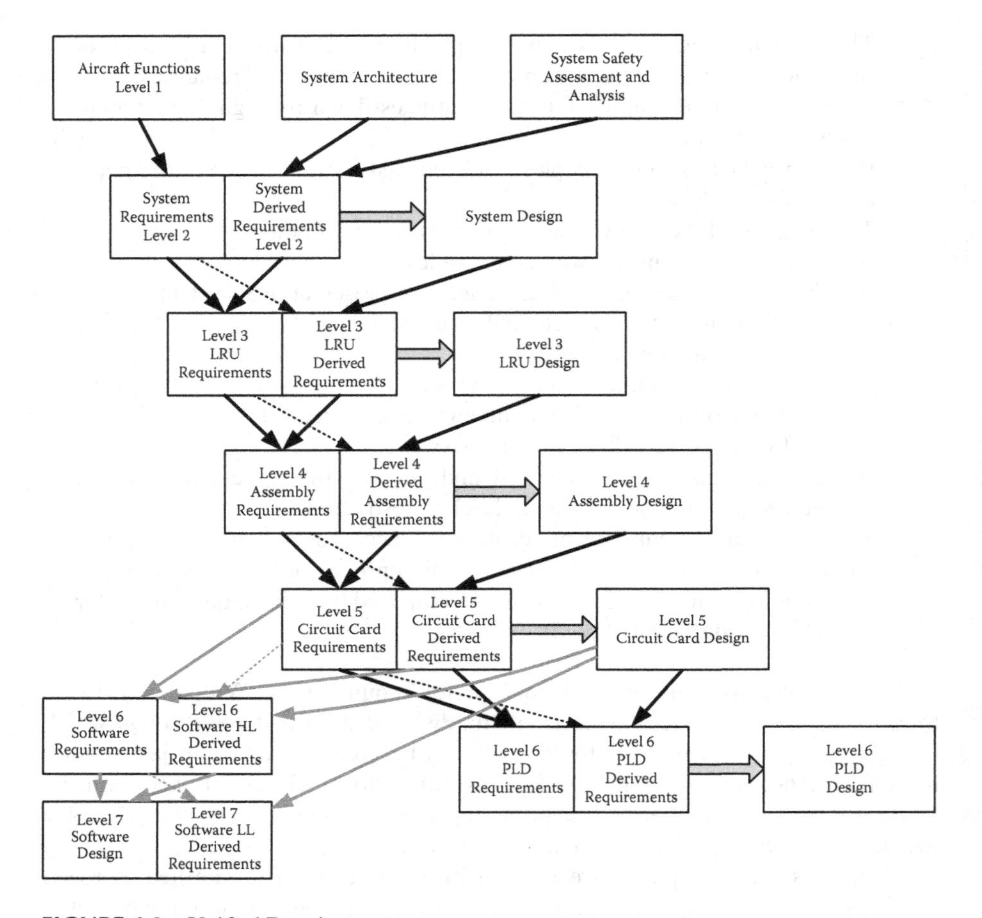

FIGURE 4.9 Unified Requirements

This alignment makes the traceability of requirements through various levels easier to identify and manage. Since design decisions are captured, derived requirements and their attendant traceability through various levels are easier to identify and manage.

The verification activities can also take advantage of and benefit from the alignment of signals and unified requirements. System level tests can be used for test coverage of system level requirements and associated LRU requirements. LRU level tests can be used for test coverage of LRU level requirements and associated circuit card requirements. Circuit card level tests can be used for test coverage of circuit card level requirements and associated software and PLD requirements. When PLDs are used as a bridge between a processor and other circuitry, the software tests for DO-178 verification credit can also be used to drive signals to/from the PLD. This allows test coverage of software and aspects of the PLD requirements to be verified together.

Currently, many projects use completely separate test teams and activities for verification of system, LRU, software, and PLD requirements. The separate approach results in duplication of effort and increases project costs. The separate approach also tends to test software and PLDs in isolation, which can miss the opportunity to observe and detect unintended functions or erroneous side effects of software and circuits. An integrated test and verification approach based on unified requirements can be used to reduce the overall number of test cases, test procedures, test results, and associated peer reviews.

5 Validation

Validation and derived requirements are intimately related: the purpose of validation is to show that derived requirements are correct and complete, and the purpose of identifying derived requirements is to identify which requirements must be validated. Thus a discussion of either topic must by necessity discuss the other, since neither can exist in isolation.

DERIVED REQUIREMENTS

The DO-254 definition of a derived requirement can be paraphrased as a requirement that is created as the result of the hardware design process. This definition is somewhat ambiguous, which often leads to confusion and questions: How does the design process create derived requirements? What are the characteristics of derived requirements? How are derived requirements identified and differentiated from non-derived requirements?

Derived requirements are created when functions are added to the hardware to support design decisions, when higher level functions are decomposed as they flow down to lower levels, and when a function is created that does not support a higher level function. The common characteristic of all of these requirements is that they contain new information that must be assessed—in other words validated—to confirm that the new information is correct, complete, and consistent with the intended functionality of the system (as expressed by system requirements).

In the software domain, Section 5 of DO-178B states that a derived requirement is a requirement that does not directly trace to a higher level requirement. This differs significantly from the definition in DO-254. As stated in previous chapters, the inertia of the initial DO-178B connotation of a derived requirement often finds its way into the hardware domain. This is understandable given that for many years the DO-178B definition and usage was the only one available. Considering the widespread use of the DO-178B definition and usage and its potential to create problems with hardware, it is appropriate to discuss that definition and how it differs from the DO-254 definition in both context and common use, and how those differences can affect the processes in DO-254.

The definition of derived requirements found in the glossary of DO-178 is almost identical to the one found in the glossary of DO-254. However, while Section 5 of DO-178B restates the definition as requirements that do not directly trace to a higher level requirement, DO-254 does not. Instead, DO-254 and its processes stick to the glossary definition, which allows for derived requirements that trace to parent requirements. In fact, instead of stating that derived requirements do not trace to higher level requirements, DO-254 does the opposite: paragraph 5.1.2-8 in DO-254 specifically states that requirements should be traceable to the next higher level of requirements, and that *derived requirements should be traced as far as possible through the hierarchical levels*. This statement leaves no doubt that hardware derived requirements can, do, and are intended to trace to higher level requirements.

The DO-254 definition is highly appropriate for hardware requirements because the decomposition of functionality (into derived requirements that trace to parent requirements) through successively lower levels of the system is one of the basic design methodologies used in hardware system design. Without this traceability from parent requirements to lower level derived requirements, tracking functionality as it decomposes from the system to the component level has considerably less meaning and loses much of its usefulness.

When hardware functionality flows downward it often gets decomposed and elaborated, and these decomposed and elaborated functions do (and should) trace to their higher level counterparts. This is a core aspect of the design of large hardware systems, and the traceability through the levels of hardware is how the implementation of the system functions is ensured and managed.

So hardware derived requirements normally trace up to a parent, or higher level, requirement as the flow down of system functionality is managed through the decomposition of the functionality captured in requirements. Under these conditions the two definitions of derived requirements will conflict and result in reduced design assurance in hardware because a significant portion of the hardware derived requirements will be misidentified as being non-derived and will therefore not be validated.

Derived requirements may also trace up to—in other words be derived from—parent hardware design features and the design decisions that created them. This is worth noting because the ability to trace the functionality captured in derived requirements is one of the most useful benefits of traceability: not only does it allow tracking and confirming system functionality through and down to the lowest levels of hardware during the design process, it also greatly facilitates the tracking and management of future changes to the design. If the DO-178 connotation of derived requirements is used indiscriminately with hardware, much of the utility of traceability is lost, leaving a significant gap in design assurance both during initial development and when making future changes.

Thus traceability is simply an aspect of functional decomposition that provides a useful means for keeping track of system functions but actually has nothing to do with whether a requirement should be validated, so using traceability to identify a derived requirement is neither effective nor sensible for hardware.

Using the DO-178 connotation of derived requirements will typically cause a conundrum: if derived requirements are defined as requirements that do not trace upward, they will lose their connection with the higher level functions that spawned them, rendering the traceability useless for tracking functions from the system level. But if the traceability is used correctly to trace functionality from the system level, the derived requirements will have to trace to their higher level functions, robbing them of their derived status and thereby eliminating them from validation. Thus if the DO-178 connotation of a derived requirement is used with the DO-254 development processes, there is a danger that the validation process will be undermined or even break down altogether. At the very least it can mean a choice has to be made between having effective validation or effective traceability.

One way to get beyond this conflict of definitions and plug the resulting holes in hardware design assurance is to find a definition for derived requirements that does not rely on characteristics that are irrelevant or even contrary to the goals and objectives of DO-254 and its processes. As stated earlier, validation and derived requirements are intimately related, so going back to the objectives of the validation process in DO-254 can help understand what a derived requirement really is, and in the process "derive" a definition that is both sensible and practical.

The objective of validation is defined in DO-254 as verifying that derived requirements are correct and complete. In the DO-254 realm, only derived requirements are validated so the point of identifying requirements as being derived is to identify the requirements that should be validated for completeness and correctness, and they should be validated because they contain some kind of information that must be verified for integrity. If the validation objective is turned around, a suitable (and in many respects more useful) definition of a derived requirement is *a requirement that needs to be validated to ensure that it is correct and complete.* This definition can be refined to state that *a derived requirement is any requirement that contains information that must be validated.*

The above definition has some advantages over both the glossary definition and the DO-178 connotation: it is more easily applied than the glossary definition, it is independent of traceability (as it should be), and it is derived from and therefore satisfies the objectives of DO-254 and validation in particular.

CREATING DERIVED REQUIREMENTS

At the aircraft level, all functions (again, expressed as requirements) are originated and thus those requirements need to be validated. The validation of these requirements ensures that this starting set of requirements is complete and correct, ensuring that the aircraft function starts with the right set of requirements that fully specifies the functionality from the aircraft level perspective.

At the next lower level of the system, some of the aircraft level functions (i.e., requirements) are flowed down unchanged (allocated) if the functionality in those requirements is already appropriate for the next lower level. Some of the functions (and their requirements) are decomposed to express their constituent functionality in terms that are appropriate for the next lower level, creating derived requirements in the process, and new functions that are not directly related to the higher level requirements are added as well, creating more derived requirements. Higher level design features will also dictate the functionality of lower level hardware, normally to define the interaction between the higher and lower level hardware, or integration of the lower level hardware into the higher level hardware, creating more derived requirements. This process is repeated at every level of the system. **Figure 5.1** shows the requirements decomposition and allocation, and addition of derived requirements.

Derived requirements can originate or add functionality at all levels of abstraction or design. These derived requirements need to be validated at the level at which they

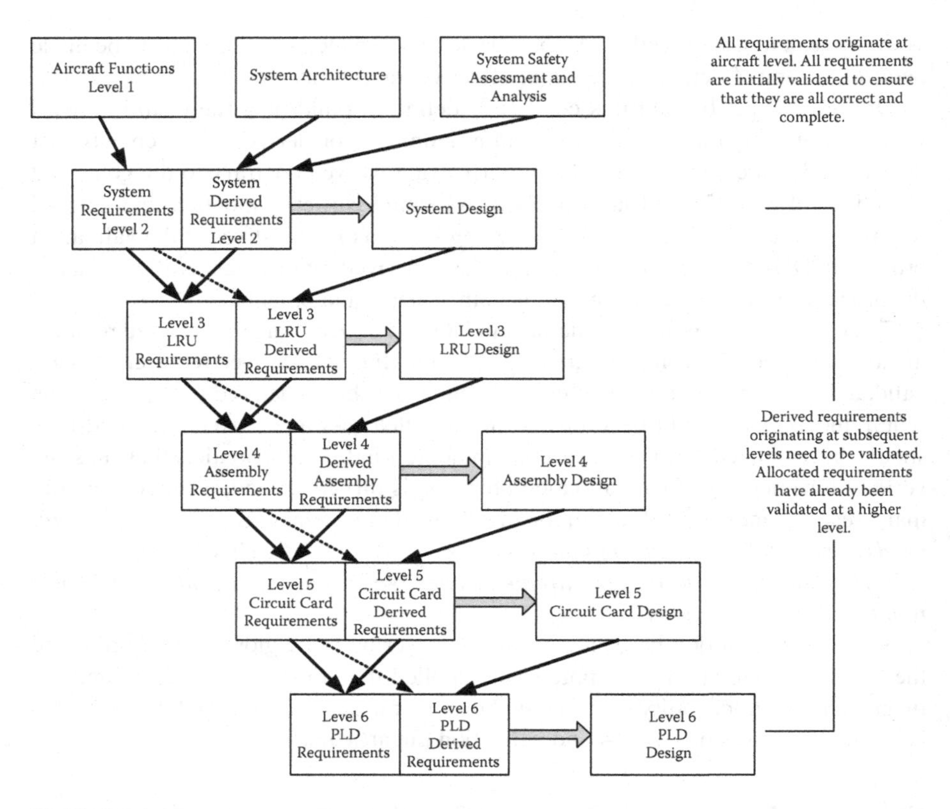

FIGURE 5.1 Requirements Flow Down

were created to ensure that they are correct, complete, consistent with higher level functionality, and appropriate for their level in the system.

Derived requirements must also be assessed from the system and aircraft perspective to ensure that there is no safety impact. The worst case scenario would be if a function is added to the hardware that is in a higher level functional failure path.

Derived requirements should be identified as being derived, and as with non-derived requirements they should also be flagged if they are related to system safety. Derived requirements should also include a justification, or in other words validation data, that is used to justify the requirement and its parameters. Ideally this justification will include enough information to allow a reviewer to determine whether the derived requirement is both correct and complete. Validation of the requirement should include an analysis of the justification to make that determination. If the reviewer cannot determine that the requirement is correct and complete through an analysis of the justification, then it is normally a sign that the justification is incomplete.

Even arbitrary specifications should include a justification that explains how the arbitrary value was selected. For example, if an FPGA output drives the clock input of an ADC, and the idle value of the clock (the logic level of the clock when a clock signal is not being generated) can be either high or low without affecting the functionality of the ADC, the requirements can arbitrarily specify that the idle value be set to one value or the other when the clock is not being generated. While the specified logic level may be arbitrary, it still should be substantiated by describing the justification for the arbitrariness of the decision. In this case, the justification for the requirement would be that the particular ADC being driven by the clock can accept either a high or a low for the idle value, so it was arbitrarily set to one value. In the case of arbitrary specifications, the very fact that the specification is arbitrary is important information that needs to be documented so that future changes to the design can take into account that the specified value can be changed without affecting functionality. If the arbitrariness of the design is not documented, engineers working on the system in the future, or using the FPGA as previously developed hardware in a new application, will not have all of the pertinent information at their disposal and therefore will not be in a position to make fully informed decisions.

VALIDATION METHODS

While validation can be achieved through review, analysis, or even test, the most common method to validate derived requirements for airborne electronic hardware is review. The review of derived requirements can be performed as a separate activity with its own criteria and results, or may be combined with the review of other non-derived requirements, as long as the review criteria are clear about which type of requirements they apply to. In some cases, it may be easier to review all requirements, derived or not, with all review criteria. An advantage of this approach is that it will assess allocated (directly flowed down) requirements for their applicability and appropriateness for the current level of the system, ensuring that all functions and the requirements that express them are appropriate for that level of hardware. Ignoring allocated functionality and requirements will not confirm that they are appropriate, possibly resulting in requirements that are inappropriate for the hardware level at which they exist.

The review criteria for validation should include a check that:

- Each derived requirement is correct
- Each derived requirement is complete
- The justification provides a rationale for the derived requirement
- Each derived requirement has been assessed for impact on system function
- Each derived requirement has been assessed for impact on aircraft function
- Each derived requirement was properly identified as a derived requirement
- All non-derived requirements were properly identified as non-derived

Reviewers performing the validation against the derived requirement review criteria should include team members representing:

- The author of the requirements for the end item (PLD or circuit requirements)
- The designer of the parent hardware—especially when the derived requirement stems from a design decision
- The system requirements

- The aircraft requirements
- Safety
- Designee or certification authority for airborne electronic hardware
- Designee or certification authority for the system

When a derived requirement is added for support of a design decision, for example if PLD requirements for how the PLD should interface to an analog to digital converter (ADC) device are derived from the design decision to use the PLD to control a particular ADC device, the review should assess the derived requirements against the design data of the upper level design. For a PLD, this could translate into reviewing PLD requirements alongside data sheets for integrated circuits that are peripheral to the PLD. If the circuit card designer decides to use a specific circuit that the PLD drives, then the PLD outputs need to meet the signal levels and timing specified for that circuit.

When a derived requirement is added to support manufacturing or test, then the derived requirements need to be reviewed against the specification of the manufacturing or test equipment.

Analysis is used extensively in validation. Even a requirements review will require some amount of analysis to determine whether a derived requirement is correct and complete, so in reality review and analysis often go hand in hand. If a derived requirement contains a translation or conversion of performance parameters, such as converting a hydraulic actuator displacement parameter at the system level to its equivalent voltage or current control signal output at the electronic box level, analysis can be used to confirm that the conversion from mechanical to electrical parameters was accurate. Note that test can also be used to validate the conversion if a suitable test setup is available; in fact, test is arguably the preferred method since it provides a real world conversion between mechanical and electrical parameters.

Test can also be used to validate requirements that specify design features (implementation) rather than functionality. As discussed in the verification chapter, implementation requirements can unintentionally interfere with the independence of verification, and also prevent verification from showing that the hardware implements its intended functionality because the design rather than the intended functionality is captured in the requirements. When this happens the burden of proving intended functionality falls upon validation, and the only real way to show that the hardware is meeting its intended functionality is to test the hardware in much the same way that verification would. Validation through test can be conducted on prototype hardware to show that the design being specified will perform as intended. Validation through testing of finished hardware will require that the hardware be designed and built against unvalidated requirements, which can result in rework if the requirements were in error.

6 Philosophy 101—Design Assurance Through Design Practice

As noted previously, DO-254 contains little or no technical information that an engineer can use to create safe and reliable electronics, focusing instead on the processes and methodologies that will first ensure that minimal errors are inadvertently inserted into a design, and second, ensure that the errors that are inserted can be revealed and fixed. The absence of technical guidance is appropriate since codifying the technical basis or standards for a field as rapidly evolving as electronics would seriously cripple the avionics industry in a very short time. Thus the technical aspects of a design are predominantly left to the discretion of the design engineers and their company design standards, subject to the technical review and approval of FAA regulators and their designated individuals or organizations. So while the technical content of a design may not be guided by DO-254, there are other checks in place to ostensibly ensure that the technical content complies with the technical goals of the FARs.

Some people may find this aspect of design assurance rather disconcerting, but it is worth noting that if the processes used by the designers comply with the objectives in DO-254 and are executed in a conscientious manner, much of the uncertainty in the design will be taken out of the equation. Eliminating this uncertainty is an unstated objective of DO-254.

DO-254 defines design assurance as a methodology for identifying and correcting errors as a way to satisfy the regulatory requirements. In many respects the DO-254 definition summarizes much of the content of DO-254, including the lack of specific technical guidance.

The DO-254 definition of design assurance, when considered against the entire scope of high reliability system design, addresses what could be considered the "back end" of design assurance. It is essentially reactive in that it focuses mainly on the quantitative process of detecting and eliminating errors (as opposed to the mostly subjective art of identifying error-free but unsafe design features) after they have occurred or been otherwise introduced into the design. It does little to directly address the "front end" of design assurance, which is to prevent unsafe design features that are either intentionally or unintentionally designed into the hardware by engineers or design tools. DO-254 does provide for design reviews that can be used to detect unsafe design features, but it relies heavily upon reviewers who are competent to identify and understand unsafe features, which may not be possible if a company's engineering culture is permeated by imperfect design methods and philosophies. In other words, if a company's engineering culture promotes and standardizes the use of unsafe design features and methods, the reviewer's perspective will reflect that culture and legitimize those features and methods, so peer reviews will only confirm that those unsafe design features and methods are "properly" in the design.

Design assurance as presented in DO-254 is implemented through multiple means at multiple points in the design process, as well as at various levels of project execution. In addition to the obvious aspect of design assurance that focuses on a well-planned and systematically executed design process, there are supporting processes that work in the background to continuously confirm that the design process is producing the correct output at the correct time and with the correct integrity. These supporting processes consist of configuration management, process assurance, validation, verification, and certification liaison, all of which are peripheral to the design process but are essential if it is going to generate designs that can satisfy the FARs.

The interaction between the design process and the supporting processes can be considered analogous to the architectural mitigation techniques described in ARP4754 and in Appendix B of DO-254. As noted in the Introduction to DO-254 chapter of this book, architectural mitigation is the means by which electronic circuits can, through high-level architectural design methods, realize a system reliability that is orders of magnitude higher than the inherent reliability of the electronic components themselves. Likewise, the use of a similar philosophy in a process system can result in a level of design assurance that exceeds what is possible just through the application of a structured design process. In this case the design process, like electronic circuits, can only realize a limited level of excellence on its own. It is the supporting processes—analogous to architectural mitigation—that enable the process driven methods to identify, capture, and eliminate design errors and thereby increase the effective reliability of the process by many orders of magnitude, similar to how architectural mitigation captures, isolates, and nullifies failures and errors generated by electronic circuits.

The design process that is presented in Section 5 of DO-254 is an example of the "classic" design process. Classic design is the process that is often taught in basic engineering courses to introduce students to the concept of following a structured process to ensure that their designs are executed in a logical, systematic, and efficient manner that maximizes control over the design and minimizes the chance of errors. All projects, regardless of their size, complexity, or goal, will cycle through the phases of the design process even if the designers do not realize it and are not consciously trying to follow it. This is because the design process mirrors the thought processes that engineers will naturally employ when solving a problem: figure out what is wrong, decide what the fix or solution has to do if it is going to fix the problem, think of a way to implement the fix, create the fix, and then test the fix to make sure it actually fixes the original problem. The process is logical, and in the long run it is the shortest path to the goal.

The design process in Section 5 of DO-254 consists of five "phases" of design activity:

1. Requirements Capture Phase, where the hardware item's requirements are conceived, written, captured in documents and/or a requirements management system, and version controlled in a configuration management system as appropriate for the hardware's DAL.
2. Conceptual Design Phase, where the high level strategy for implementing the functionality expressed in the requirements is conceptualized and documented.
3. Detailed Design Phase, where the conceptual design is elaborated and refined into the design (usually HDL code for PLDs and schematic diagrams for non-PLD electronic hardware) that will be implemented in hardware.
4. Implementation Phase, where the detailed design is converted to its hardware implementation and then development tested to ensure that the hardware works as designed.
5. Production Transition Phase, where the final version of the hardware design is readied for series production.

In theory these phases are intended to be executed in order, but in reality the practical considerations of project schedule and resources will often require that some of the phases overlap or even run concurrently. While this is not considered strictly "correct," the process is flexible enough to accommodate such variations and still produce adequate design assurance. If an organization's normal design practices include such overlaps and concurrency, the design process description in the project HDP should make a note of this along with a justification that substantiates the claim that such variations can be tolerated without sacrificing or impacting the design assurance of the process and its resulting hardware item.

The design process described in the HDP should also account for the continual interaction with the supporting processes, in particular the configuration management, validation, and verification processes.

The design process that is used (and documented in the HDP) does not have to be the same as the design process in DO-254, but it does have to satisfy the objectives that are stated in paragraphs 5.1.1, 5.2.1, 5.3.1, 5.4.1, and 5.5.1. Virtually any design process can be acceptable as long as the process can be shown to fully satisfy those objectives and generate the artifacts that will substantiate that those objectives were met. The processes also need not have the same phases as long as the phases that do exist support the DO-254 objectives. This "mapping" of a different design process to the process objectives in DO-254 is generally straightforward because of the way that the DO-254 process captures the natural flow of a design project from conception to implementation: as the adage goes, a rose by any other name is still a rose, so attaching different labels and phases to a design process will not change the essential flow of development as long as the process has reasonable integrity and does not contain too many unusual or excessively arcane practices.

Note that the design process does not have to satisfy the activities that are documented in paragraphs 5.1.2, 5.2.2, 5.3.2, 5.4.2, and 5.5.2 of DO-254. The activities listed in those paragraphs are suggestions for how the objectives can be satisfied, but are not mandatory aspects of the design process, and are not required for a design process to be DO-254 compliant. However, most design processes with the requisite integrity for AEH design will incorporate most if not all of the guidance contained in those paragraphs.

DO-254 Section 5 also contains subsections on acceptance testing (5.6) and series production (5.7), but neither of these topics are significant factors in the design and development processes.

A DO-254 compliant design process differs from the classic design process in that it has supporting processes to boost the integrity of the basic design process and make it more effective. The design process has a requirements capture phase in which the solution's requirements are developed and recorded, and the validation supporting process scrutinizes those requirements to make sure they are correct and complete, ensuring that errors introduced during that phase are minimized or eliminated. The design is created during the conceptual and detailed design phases, and the verification supporting process scrutinizes the design to make sure it is correct and complete, ensuring that errors introduced during those phases are minimized or eliminated. When the hardware is created during the implementation phase, the verification supporting process steps in again to scrutinize the hardware and conduct detailed tests on it, ensuring that the hardware is correct and complete, and that errors introduced in that phase are minimized or eliminated.

Throughout these phases, the configuration management supporting process creates an environment where each version of the requirements, design, and verification data is documented, identified, and managed, ensuring that errors will not be introduced through mishandled data. At the same time, the process assurance supporting process scrutinizes the data generated by the design and supporting processes to ensure that everything was performed when and how they were supposed to be performed, and the certification liaison supporting process scrutinizes the entire project to make sure the design and supporting processes are being implemented and executed properly.

The end result is that the supporting processes obtain every bit of effectiveness from the design process to ensure that any hardware coming out of it has as much integrity as it possibly can.

However, as mentioned previously, while the design and supporting processes can effectively minimize the errors that are introduced into the design and maximize the errors that are uncovered and fixed, it is not designed to identify and fix design features that are inherently unsafe or can adversely affect the integrity of the hardware. Identifying and fixing weak or inherently unreliable design features requires qualitative judgment backed by experience, knowledge, and technical competence—things that cannot be acquired from a process or its governing document. On the other hand, developing those technical skills can be fostered or synthesized through adopting and adhering to a good set of design standards and philosophies that can be used within a process to minimize or eliminate the introduction of unsafe or unreliable design features.

DATDP

Design Assurance Through Design Practice (DATDP) is an engineering approach that emphasizes the use of a good design philosophy to engender solid engineering practices and methodologies to promote safe and reliable designs. It is applied at the circuit design level and is, by and large, the result of lessons learned from decades of experience in designing (and auditing) high reliability electronic systems. Fundamentally its approach is to use robust design techniques and philosophies to design reliability in, rather than rely solely on testing errors out. It is the process of proactively designing in reliability instead of reactively weeding out errors.

Like DO-254, DATDP does not prescribe specific design techniques or circuit architectures, although the design standards that may come out of DATDP might. Instead its focus is on developing a mentality that enables engineers to create designs that are safe, and then to objectively examine those designs from all angles to find potential weaknesses that could compromise the integrity of the system.

DATDP addresses three aspects of front-end design assurance: device selection, design philosophy, and design execution.

Device Selection

Device selection focuses on the components that are used in a design. While most components will be adequate from the reliability perspective if they meet the environmental requirements of the system, there are still non-environmental considerations that can affect a component's ability to support the necessary reliability for airborne systems.

Some recommended device selection guidance includes the following. The guidance in this book is written to focus on programmable logic devices, since that is where the current application of DO-254 is focused, but can be applied to other components as well.

- **Let the system select the devices.** In other words, let the system's functional, reliability, safety, cost, and related requirements, not personal preference, guide the component selection process. It is acceptable to have favorite components, but their use should be dictated by the system requirements.
- **Power requirements, sequencing, and consumption versus temperature.** Beyond the basic parameter of power consumption are the larger scale considerations for the range of power supplies, how the power supplies must be managed, and how the power consumption may vary with temperature. Keep a more global view and consider how a device may affect the complexity and cost of the rest of the system. For example, an FPGA that requires two power supplies that must be sequenced in a specific way may not be a good choice if its power supply and management circuits add appreciably to the complexity of the system. In addition, some devices that consume very small amounts of power at room temperature can become considerably more power consuming at higher temperatures. If a system's operating temperature range overlaps the temperature at which the device consumes more power, the device may not operate at as low a power level as originally intended, and in some cases may be susceptible to thermal runaway. Research power considerations thoroughly before selecting any device.

- **Radiation tolerance.** Although radiation is not normally a consideration for commercial aircraft, radiation tolerance includes susceptibility to single event upset (SEU) events, which are relatively common at the normal cruising altitude of commercial aircraft. Different semiconductor device technologies have different levels of sensitivity and susceptibility to SEUs, so keep SEUs in mind when selecting devices.
- **Service life.** Do not sabotage a system by selecting a part that may go out of production during the life of the system. Look for parts that are mature (have been in large scale use long enough to establish their trustworthiness) but not likely to be discontinued anytime soon. Also look for manufacturers that ensure availability of suitable parts for medical and automotive markets.
- **Lead time.** Designing a device into a circuit and then finding out that the real devices will not be available in time for your project can cause more than schedule delays, it can result in design errors when a new device type is designed in as a replacement. Be aware of a device's availability and lead time so that the designer will not be surprised.
- **Technical support.** Not all device vendors are equal in this regard. Working with a company that has fast and reliable telephone technical support can be a real time (and cost) saver when schedules get tight and problems arise.
- **Product support.** Implementing a design in a given device is the other half of creating hardware. Are the vendor's design tools easy to use and understand? Is the user interface logical, and do the tools have features that can introduce errors into the design?
- **Packages.** Does the selected device come in the type of package that best serves the needs of the system? Some package types are better suited to harsh environmental conditions than other types. Think about the environmental and electrical conditions of the system when selecting a device and its package. Device storage and handling aspects can also affect package choices. Also consider access to device pins for in-circuit testing and whether suitable sockets are available to facilitate testing.
- **Device features.** Integrated circuits have been evolving for decades, and with each decade the devices get considerably more capable, with an attendant increase in complexity. FPGAs in particular come packed with more and more features ranging from mathematical resources to built-in processors. This adds enormously to the capabilities of the device, but with the capabilities comes complexity, and with complexity comes an increase in the downstream burden of verification. In addition, if the design does not use all of those features, some certification authorities may express concern over these unused functions and how the design can ensure that they will not present a safety problem.
- **Service history.** Using new devices, no matter how capable and ideal they are for a design, is not always the best way to go. Devices that have no commercial or industry track record can be problematic with the certification authorities. Before selecting such a device for a design, discuss it with the certification authorities to get their concurrence.

- **Semiconductor technology.** Different semiconductor materials and fabrication technologies will have different robustness characteristics, and some types of semiconductors that perform flawlessly in a consumer product may have weaknesses that could preclude their use in aircraft systems. Before selecting a device, conduct an analysis to determine whether its semiconductor material, architecture, feature size, gate type, and programming methods can affect the safety of a system when subjected to the conditions aboard an aircraft in all of its potential operating environments.
- **SEUs.** Single event upset (SEUs) considerations are related to semiconductor technology because susceptibility to SEUs can be highly influenced by it. Some types of devices are simply more susceptible to this error source. Soft errors will affect all technologies more or less equally and can be mitigated through system level design features such as architectural mitigation or simply refreshing data on a periodic basis. Hard errors, which can change the programming of a PLD, are a more serious issue and are dependent on the semiconductor technology.
- **PLD size.** The size of a PLD can affect more than just how many circuits can be designed into it. In some cases, a device that is significantly oversized for its application can raise questions and concerns from the certification authorities about the disposition of the unused resources. Size a PLD to fit its application while allowing room for the design to grow, but keep it within reason.
- **Speed.** Faster, like bigger, is not always better. Faster devices can mean higher susceptibility to noise and errant signals, more cross-talk and reflections at the circuit card level, and more radiated noise.
- **Data retention in flash devices.** Flash-based PLDs have a limited data retention period, and that period may depend on temperature. Do not rely solely on the data retention banner on the first page of a data sheet; study the data retention tables in the data sheet to get a more realistic value. It may be that the advertised 20 or more year data retention period could fall considerably if the device is operated at high temperatures.
- **Power-on performance.** Some devices have to load their configuration from external memory each time they power up. If the configuration time is greater than the specified start-up time or required availability for the system, then it may not be wise to select that part.

Design Philosophy

Design philosophy employs mental attitude, thought processes, and design rules that guide the design process and provide a sound basis for creating safe and reliable designs. Design philosophies are used to create a mindset and attitude that are conducive to high integrity designs.

The heart of DATDP is a set of rules that together comprise a design philosophy that has served well as the basis for creating safe and reliable designs. These rules, facetiously called "Roy's Rules," are at times frivolous, but in their essence they embody an engineering mindset which, when applied to a design, can make the difference between a reliable, high integrity system and a weak or marginal one.

Roy's Rule #1: Passing the Buck Is Expensive (or Carry Your Own Burden)

Experience has demonstrated that the later a problem is fixed, the harder and more expensive the fix becomes. Similarly, passing a task to the downstream user of a product can have a similar effect, and increase the likelihood of errors as well. For example, if a requirement author does not want to go through the effort of writing down the fine details of a specification and elects to let the reader of the document "figure it out themselves," the requirement has just gotten considerably more expensive, and the chance of errors has gone up as well. Considering that a document (or in this case a requirement) will typically have one author and many readers, and the author knows the topic best and can provide the details more efficiently than anyone else whereas the readers will have less expertise and will take considerably more time to generate the same information, then letting the readers derive the details for themselves is a bad business decision at best, and could be a problematic one as well since the reader is more likely to make a mistake when deriving that information. Supposing that a requirement author could write down the details of one requirement in five minutes whereas a reader might take ten minutes to figure it out on their own, and if there are ten readers who will use that requirement, then the author's failure to provide the detailed information has increased the cost of that requirement by a factor of 20. If that figure is multiplied by the number of incomplete requirements in a document (which is likely to be significant given the attitude of the author), then it quickly becomes clear that the simple act of delegating the details to the reader can be a very costly proposition, and that does not even include the cost and time associated with fixing errors that may occur.

It is tempting for a requirements author to leave some details to the readers to figure out, normally as a way to avoid spending time on something that the author may feel the readers can figure out on their own. In other cases, the author does not realize that their knowledge or experience is not common or universal. As described earlier, however, delegating the details in this way can result in wasted time and effort, delays due to mistakes in figuring out the requirement details, delays and errors when the requirements are used for verification (even more if the verification engineer misinterprets the requirement and creates erroneous test cases), and expensive hardware fixes if the mistaken interpretations are not caught early and find their way into hardware. DO-254 lists completeness as one of the objectives of validation, which means that a derived requirement that does not provide all relevant details is insufficient and should be corrected.

Another area where there is a strong temptation to delegate work is writing comments in HDL source code. Again, the reason is typically that a designer feels (with good intention) that some level of passing effort to the reader is harmless and will save time and effort on their part, not realizing that it will multiply the time and effort expended by anyone who reads the code, such as by a reviewer during a code review, when making changes later in the project, or when reusing the code at a later date in a new application. And as with requirements, poorly commented code will require that the reader interpret and figure out the functionality, which means there are now opportunities for misinterpretations and outright mistakes, which can result in more serious delays and costs if the mistakes find their way into the final hardware. Well-commented code will often have more lines of comments than lines of code. While this seems unnecessary to some people, anyone who has had to review or work on code that was commented less than this can testify that delegating the details to the reader is not the way to save time and effort.

Roy's Rule #2: Predictability May Be Boring in People But It Is Goodness in Electronics

Deterministic operation, where there are no variations or uncertainty in how a system operates each time it operates, is one of those behavioral traits that can be immensely boring when exhibited by a person, but highly attractive in an electronic system. A deterministic system will behave predictably under all operating conditions, and is essentially incapable of changing its behavior regardless of the inputs it receives from the rest of the system. When a system or circuit is designed to operate in this manner, it is inherently immune to outside influences, including abnormal inputs. However, it is not the deterministic behavior that creates this immunity; instead, the determinism and immunity are both characteristics of a type of system or circuit that controls or is independent of, rather than reacts to, its environment. Or alternatively, a system will exhibit deterministic behavior if its inputs define its data output but not its operation and behavior.

A very simple example of this concept is shown in **Figure 6.1** through **Figure 6.3**. **Figure 6.1** shows a very simple finite state machine (FSM) that implements a control interface for an analog to digital converter (ADC) device. It is typical of how

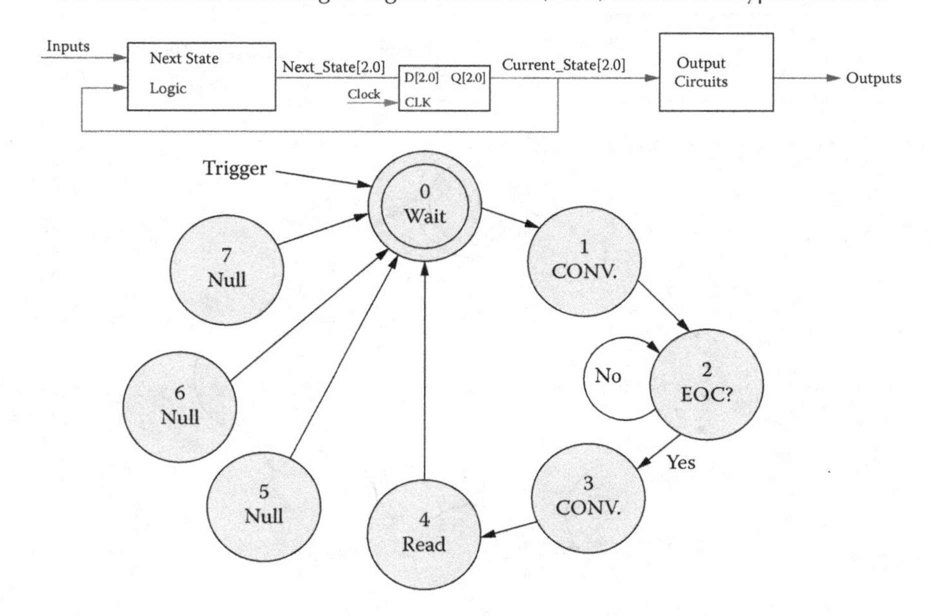

FIGURE 6.1 Example One—Typical FSM Control Circuit

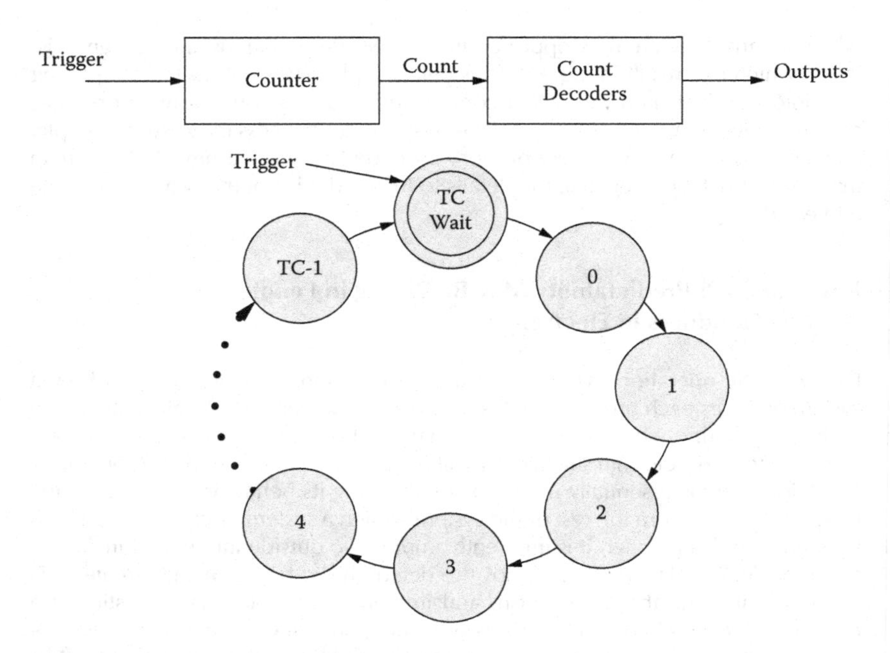

FIGURE 6.2 Example Two—Deterministic Circuit

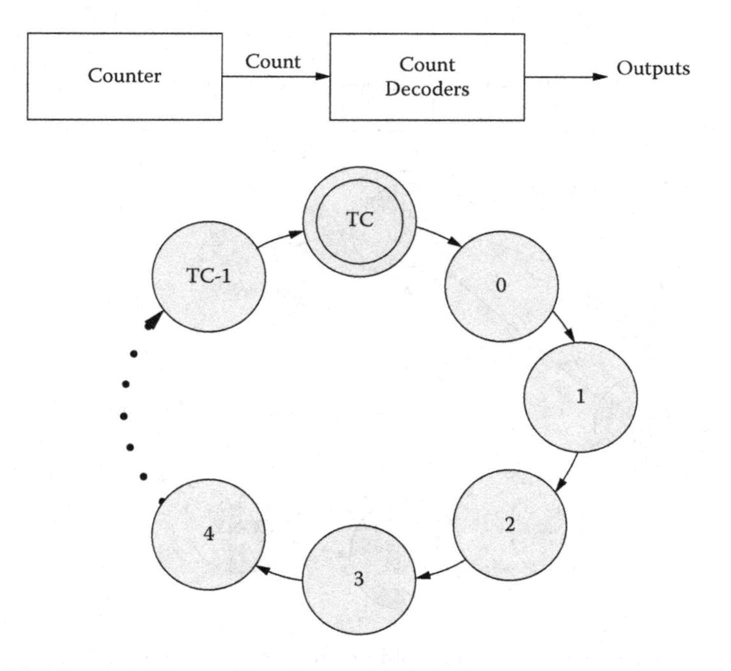

FIGURE 6.3 Example Three—More Deterministic Circuit

control circuits are implemented, particularly for devices that provide feedback. In this case, the state machine holds in an idle state until it receives a convert trigger, at which point it increments to state 1 to start a conversion cycle in the ADC, and then increments to state 2 to wait for the ADC's end of conversion (EOC) signal. When the EOC signal is received, the state machine switches to state 3 to assert the ADC read control signals, and then to state 4 to read the ADC's output data before going back back to state 0 to await the next conversion trigger.

An examination of this implementation to identify potential failure modes will immediately zero in on state 2 because the state machine could latch in that state if the end of conversion signal does not arrive as expected. There are also three unused states (states five, six, and seven) that could become undefined states if not properly managed in both the source code and the synthesis tool.

An improvement on this circuit is shown in **Figure 6.2**. The circuit in **Figure 6.2** employs a counter in place of a state machine. The counter is reset to zero by the conversion start trigger, and then increments from zero to its terminal count, where it then latches and awaits the next conversion trigger. The control signals for the ADC are generated by decoding the counts. This circuit is semi-deterministic in that it does not have an equivalent to state 2 in the previous example, nor does it contain unused states, but it does include a trigger input that resets the counter to zero (its starting state).

One significant difference between these two examples is that the second circuit, if using the same type of asynchronous ADC as the first circuit, will count through the ADC's maximum or worst-case conversion period (taken from the data sheet) and then read the output data rather than wait for the EOC signal from the ADC, and thus operate the ADC in open-loop fashion. However, for this type of circuit a better selection would be a synchronous ADC in which the conversion and its output are controlled explicitly by the control signals generated by this circuit.

For an asynchronous ADC the use of count states to mark time ensures that the circuit will not hang up if the EOC fails to appear for whatever reason. Thus the effect of a failed ADC or interrupted EOC signal will at most result in null data, a missed sample, or invalid data, which can be managed at the system level, and will not affect the circuit's operation in any way. This approach creates a distinct division between the circuit's (and system's) operational and data characteristics, in particular ensuring that the circuit will operate normally and deterministically regardless of whether the data or its source becomes corrupted or otherwise untenable.

The most deterministic, and thereby robust, of these example circuits is shown in **Figure 6.3**. This circuit improves on example two by removing the trigger input and turning the circuit into a free running counter that operates independently of all external signals (except the clock). In all other respects this circuit operates identically to example two.

This circuit approach works best in a master role where the counter establishes the periodicity or frame rate for the system, so it would most commonly be used at a higher level than shown in the example. In this type of system every function and circuit is operated and timed by decoding the counts generated by the central counter, ensuring that every aspect of the system is precisely timed with respect to overall operation and to each other. Systems designed with a central free running counter of this type will exhibit truly deterministic behavior where every node, register, and even data value can be precisely predicted and modeled at any point after the system starts operating.

Roy's Rule #3: Use HDL to Design Logic, Not Behavior

Of the three levels of HDL design (behavioral, RTL, and structural), register-transfer level (RTL) is most compatible with the objectives and processes in DO-254. RTL designs are compatible with a pin level description of requirements and the respective pin level testing. Structural HDL, because of its very low level of design expression, will increase the cost of a program by requiring that the processes and methodologies in DO-254 be applied at a sub-functional level. Behavioral HDL, which has the advantage of allowing a PLD design to be expressed at a functional level, might save time and effort during the design phase of a program, but because the code is describing the behavior of the design rather than the hardware itself, the logic design is left to the design tools rather than the designer. Since the processes and algorithms in the design tools are not visible to the designer, there is no way other than through exhaustive testing to determine conclusively whether the circuits created by the tools are in fact the circuits that the designer intended to put into his design. In addition, the high level at which the design is coded results in larger elements, which can affect the elemental analysis of the design. These loose ends represent a large unknown in design assurance.

RTL coding provides a good compromise between the two coding methods. It allows the designer to have a high level of control over the implementation of the design while keeping the size and complexity of the elements at a level equivalent to that of a circuit card, which is the level for which DO-254 was written. The added advantage of RTL coding is that its lower level of design expression allows the design to avoid many of the pitfalls that can be encountered during synthesis when the tools' problematic reduction features are engaged. An example of this phenomenon is discussed in the second example circuit in the Design Execution portion of this chapter.

Roy's Rule #4: Make Your Circuit Bulletproof Even If No One Is Shooting at It

This should really be obvious for engineers who appreciate the seriousness of level A design. It is not enough to design a circuit that works; a conscientious designer who focuses on safety and reliability will do their utmost to make sure that their circuits are bulletproof and will not malfunction or misbehave for any foreseeable operating condition (this phrase from the FARs should be familiar by now). An example of this rule is discussed in the first example circuit in the Design Execution portion of this chapter.

Roy's Rule #5: Use Top-Down Design or Prepare to Go Bottom-Up

Top-down design, where a system or function is first defined at the topmost level and then decomposed downward to the lowest levels of the design, is the only recommended methodology for defining a system's functionality and design. Bottom-up design, which defines a system or design from the bottommost elements and then attempts to work upward to define the system, is never recommended and normally results in curious designs and programmatic disasters.

However, the combination of top-down design and bottom-up implementation, where a hardware item is defined and designed from the top down and then assembled and tested from the bottom up, results in the best of both worlds.

Roy's Rule #6: Find Your Own Failure Modes

There are two places where this rule should be applied: first, finding all possible and potential failure modes of a design; and second, finding and knowing our own failure modes, or in other words, understand ourselves and our issues so that we can prevent ourselves from sabotaging our own work. The first is actually the easiest since designs are generally straightforward and can be analyzed to consider all possible sources of failure. The second, on the other hand, can be complex and difficult due to the unpredictability of human nature and experience. However, understanding ourselves to the point where we can recognize our limitations, biases, and eccentricities and how they affect our work will allow us to mitigate their negative effects.

Roy's Rule #7: Never Assume

An assumption is a decision that is based on a lack of knowledge—being fully informed means that assumptions are unnecessary. Some assumptions can be based on incomplete rather than no knowledge, but even then it is still a lack of knowledge that leads to an assumption instead of an informed decision. Making an assumption can jeopardize all downstream decisions and actions, so the best course of action is to avoid assumptions and rely as much as possible on informed decisions.

Roy's Rule #8: Do Not Ask for Trouble (Avoidance Is Better Than Mitigation)

This means that it is a better idea to avoid any kind of risky or questionable design feature than to put one in and then mitigate its effects. For example, one-hot state machines can be unreliable, so rather than design a one-hot state machine into a PLD and then find ways to mitigate their weaknesses, it is better to not use them at all and instead rely on an alternative circuit type that is more reliable. Intentionally introducing a weakness into a design is not compatible with the goals of safety critical design no matter how well the weaknesses are mitigated.

Roy's Rule #9: DO-254 Is Our Friend

DO-254 is a collection of industry best practices. They can improve the reliability of a system and even reduce development costs if applied properly, so the processes in DO-254 should be embraced, not shunned or avoided. In fact, considering that DO-254 contains best practices, anyone creating level A hardware should already be complying with DO-254.

Roy's Rule #10: Review Now or Pay Later

It is tempting to save time and money by skipping or skimping on peer reviews, or even by conducting low quality reviews. However, this will only cause errors to be overlooked, which allows them to propagate downstream and become more expensive to fix. It is far easier and less expensive to spend a little extra time and effort on a peer review to make sure the review is complete and thorough, than it is to fix any problems that get through because of an inadequate review.

Roy's Rule #11: Just Deal with It

Level A design is often difficult and expensive. Rather than fight it, it is usually a lot easier in the long run to just accept it and deal with it. The same applies to complying with DO-254: the most expensive approach to DO-254 is to try to avoid it, and the more you try to avoid it the more expensive it gets.

Roy's Rule #12: Ignore the Trees

Sometimes, especially when a project is behind and time is critical, the need to focus on the immediate task can make it hard to see the big picture. When making decisions, always remember to look at the impact of every decision on the long-term conduct of a project and not just on the short-term cost or benefit. What might work well, look attractive, or solve an immediate problem may not work well or could even cause problems in the long run.

Roy's Rule #13: Require Requirements

As described elsewhere in this book, DO-254's processes focus on functions and functionality, and the way functionality is defined is through requirements. And as described in the Requirements, Validation, and Verification chapters of this book, the number and quality of the requirements will have an enormous influence on the cost and effort of complying with DO-254. With requirements being as important as they are, it makes sense to invest a great deal of effort in producing high quality requirements. Experience in DO-254 has shown that requirements are consistently the single most influential data item when it comes to affecting the course of a development project, so to even consider shortchanging the requirements is inviting potential increases in development time, cost, and effort.

Requirements are not an appendage to a project, nor are they a documentation burden. They are a critical part of the design and verification processes, and it is not an exaggeration to state that the quality of the requirements will dictate the conduct of the entire project. So it behooves us to treat requirements with respect and to never try to save time and money by reducing the effort and time that is put into them.

Roy's Rule #14: Have No Faith

It can be easy to have faith in our designs, but often the reality is that our faith is misplaced or is not realistic. In other words, believing something does not make it so. Faith-based certification is not an accepted approach to compliance with the FARs; the only acceptable approach is to support all decisions and claims with hard data, so we need to know, not believe.

Roy's Rule #15: There Is No Hope

Hope is one of those terms or concepts that should never find its way into the business of Level A design. We should never hope that our designs are safe, we should only know it for a fact. If we find ourselves hoping for the best, then we probably have not done our job with diligence. In level A design there should be no hope, only certainty.

Design Execution

Design execution combines the device selection and design philosophy aspects of DATDP and applies them to the creation of individual circuits. Design execution is best presented through simple circuit examples as opposed to descriptions and concepts. The following FPGA design examples describe actual circuit issues to illustrate how design philosophy can be used to create more reliable designs and to mitigate potential error sources.

Example 6.1: Shift Registers

Shift registers are one of the simplest and most useful of the basic digital logic circuits. However, because of their simplicity and utility they often escape scrutiny and may be overlooked as a source of potential errors.

Figure 6.4 illustrates a simple three-stage shift register that shifts on the rising edge of a common clock in an FPGA. Upon examination of this circuit and application of "Roy's Rules" number 6 (find your own failure modes) to it, there is a potential error mode or weakness that is based on the fundamental criteria of how shift registers work: the hold time of each register's output must equal or exceed the following register's input setup time with respect to the rising edge of its clock input. In other words, the shift register will no longer work properly if the timing between a register's clock edge and input setup time, and the previous register's output hold time, deviate from normal specifications. These deviations can result from a number of factors, including a reduction in the output hold time of the preceding register, an increase in the input setup time, or an unequal delay in the active edge of the clock. The most common of these is a delay in the clock edge due to routing delays, which causes the clock edge to reach each register at a different time (clock skew).

Figure 6.5 shows typical waveforms for a properly functioning shift register, where the rising edge of the clock arrives simultaneously at each of the three registers. As

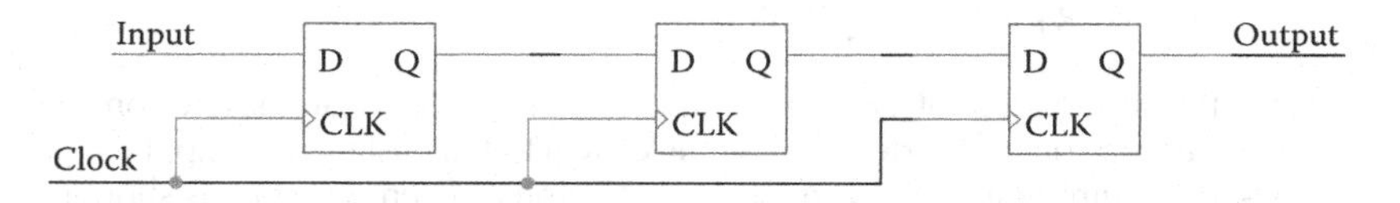

FIGURE 6.4 Simple Shift Register

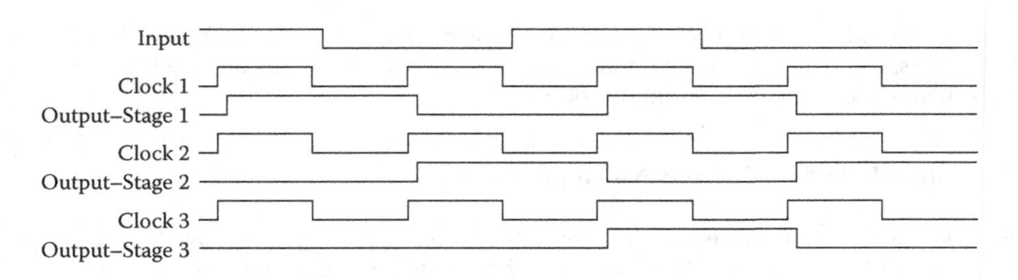

FIGURE 6.5 Waveforms for a Properly Functioning Shift Register

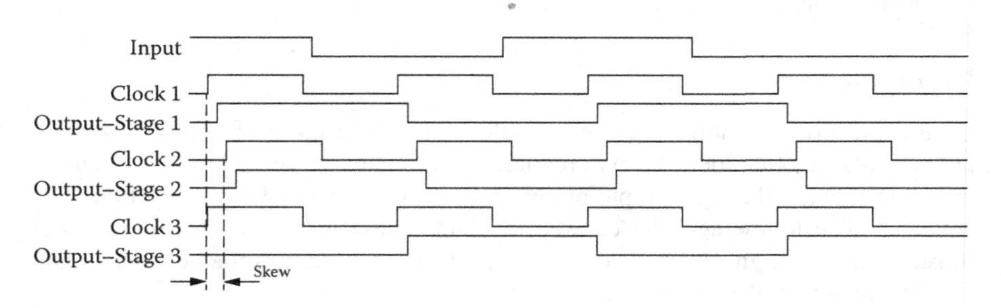

FIGURE 6.6 Shift Register with Clock Skew in Second Stage

can be seen in the waveforms, each bit of the input data stream is shifted through the three register stages and appears at the output of the shift register.

In contrast to the normally functioning shift register, **Figure 6.6** shows the waveforms for a shift register where the rising edge of the clock arrives late at the second of the three registers. Because the rising edge of the clock arrives later than the hold time of the first shift register stage, the second register will immediately shift through the data going into the first stage, causing each bit of data to shift into both the first and second stages at the same time.

Clock skew can be caused by a number of factors, such as any of the following:

1. A failure to use a dedicated low-skew clock net in the device.
2. Failure of the dedicated low-skew clock net to perform to its expectations (in actual devices these nets have been shown to occasionally still have enough skew to cause shift registers to malfunction).
3. The need to use normal routing resources because there are too many clocks for the available clock nets.
4. The place and route tool automatically removed the clock from the clock net because its assignment there conflicted with the tool's placement rules.
5. The clock could not be placed on the clock net because access to the net was determined by pin assignment, and the input clock signal was assigned to the wrong pin on the device.

Since the introduction of clock skew can have a number of causes, and some of those causes may be inadvertent or even unknown to the designer, the application of Roy's Rule number 4 (make your circuits bulletproof even if no one is shooting at it) is a good way to prevent potential clock skew problems.

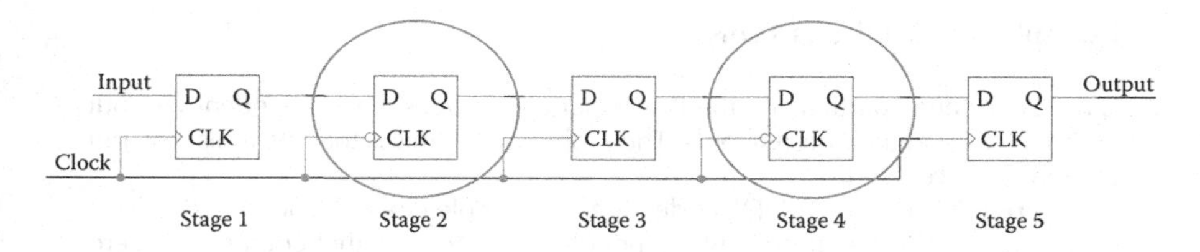

FIGURE 6.7 Shift Register with Controlled Delays

How does one make a shift register reliable? Since the problem stems from clock skew exceeding the hold time of the previous shift register stage, and assuming there is little that can be done about this skew, the logical alternative is to design a shift register that is impervious to clock skew (within realistic limits of course), and one way to do that is to force the hold time of the previous stage to exceed any expected clock skew.

Figure 6.7 illustrates one way to do this using synchronous logic. In this solution, extra registers that use the alternate phase of the clock are inserted between the shift register stages to implement controlled delays (as opposed to uncontrolled, which would be the case if asynchronous logic was added between stages) between stages of the shift register. While this may almost double the number of registers in the shift register, the cost of additional registers is insignificant in a register-rich FPGA.

Figure 6.8 shows the waveforms for this shift register. Because the newly inserted shift register stages operate off the alternate (falling) edge of the clock, the shift register will now operate correctly for any amount of clock skew up to the period from the rising to falling edge of the clock, which for a symmetrical clock will be about half the clock period.

As noted previously, the lesson from this example is not how to design a robust shift register, but rather that applying a high integrity design philosophy can predict and identify potential failure modes in even the most simple and unassuming circuits that normally escape scrutiny. Identifying and correcting these weaknesses can significantly reduce the potential for latent failures.

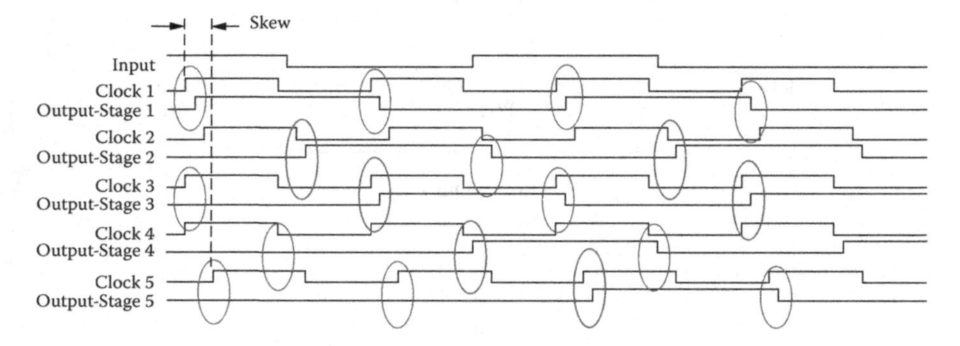

FIGURE 6.8 Waveforms for Shift Register with Controlled Delays

Example 6.2: Synthesis Tools

Circuit designers, particularly the less experienced ones, have a tendency to put too much trust in their design tools. The problem with this is that the tools may not entirely deserve that trust.

Figure 6.9 shows the VHDL code for a very simple clock divider that divides a source clock by three. It is simply a modulo-three counter that counts from zero to two and then back to zero, with count three being an unused count. There is no reset in this circuit because the synchronous device in which it is used needs its clock during reset to operate and initialize. The circuit diagram for the code is shown in **Figure 6.10**. Because the most significant bit of the counter is fed back to its synchronous clear input, both the two and three count values (binary values 10 and 11) will cause the counter to synchronously reset and restart its count sequence. Thus, if for any reason the counter reached count three (11 binary) it would immediately recover and start over.

Figure 6.11 is the logic diagram for the expected output from a synthesis tool. Its truth table shows that it should behave in the same manner as the original design.

Figure 6.12 is the logic diagram for the actual output of a synthesis tool that processed the VHDL for the clock divider, taken directly from the synthesized netlist output of the tool. A quick analysis of the synthesized output will reveal that it not only looks different than the original circuit, it also behaves differently in that it will latch in count three. This is a potential failure condition which, because the circuit generates a clock, could cause a system to lock up and fail. Hardware testing showed that the PLD would in fact lock up regularly on power up or during a power transient because the two registers in the circuit would randomly initialize to count three.

```
process(clock)
begin
if rising_edge(clock)then
 if Q(1) = '1' then Q <= "00";
 else Q <= Q + 1;
 end if;
end if;
end process;
```

FIGURE 6.9 VHDL for a Simple Clock Divider

Q(n) => Q(n + 1)

0 => 1

1 => 2

2 => 0

3 => 0

FIGURE 6.10 Logic Diagram for a Simple Clock Divider

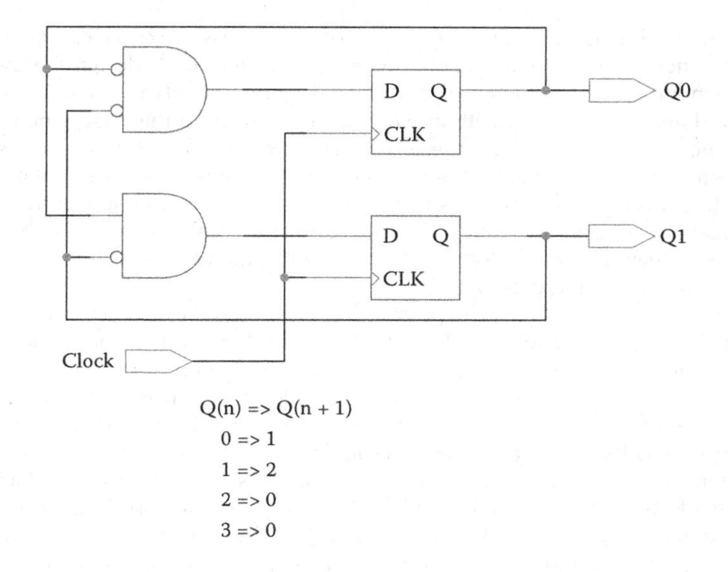

FIGURE 6.11 Expected Output of Synthesis Tool

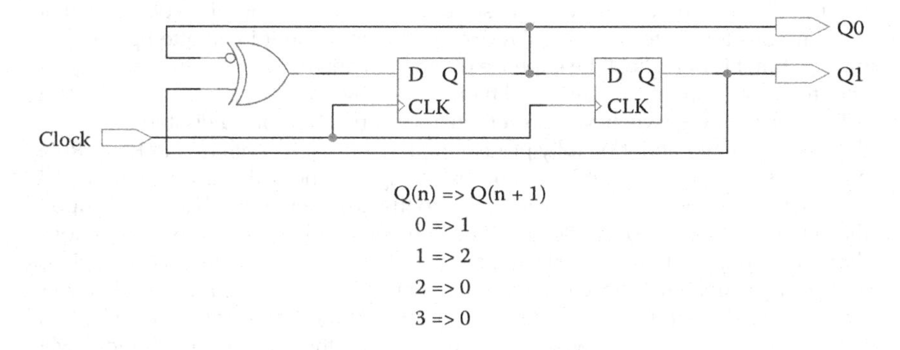

FIGURE 6.12 Actual Output of Synthesis Tool

A call to the synthesis tool manufacturer revealed that the tool was doing what it was designed to do—it recognized the circuit as a modulo three counter and intentionally changed the circuit's topology to be as efficient as possible (eliminating the unused state reduced the size of the circuit by a logic cell or two). Unfortunately it did so at the expense of changing a fail-safe unused count into a weakness that could cause a system failure. The tool manufacturer also revealed that the problematic reduction algorithm could not be turned off or defeated in that model of the tool. Since there was no software "switch" in the design tool user interface to turn off that feature, there was no way for the designer to know that that algorithm even existed, let alone that it could cause problems.

The synthesis tools used for PLD logic reduction are another example of the capability/complexity dichotomy that was brought up in the discussion of device selection, where an increase in capability will often bring with it a corresponding increase in complexity that will normally increase the uncertainty associated with using it. Since the goal of design assurance is essentially to reduce uncertainty to manageable levels, any device, tool, design feature, or design method that can increase the potential for uncertainty is at odds with this goal and needs to be dealt with in a deliberate manner. In this case, since there is no alternative to using a synthesis tool when designing with PLDs, and since the synthesis tool is an integral part of the design process, some other means of mitigating this type of effect (this is just one example of who knows how many such features exist in these highly complex but capable tools) must be brought to bear to maintain some degree of control (and therefore certainty) over the implementation of the design.

Synthesis tools are quite adept at recognizing certain code constructs and design techniques and then applying their technical magic to improve on them, often without the knowledge (or consent) of the designer. In the vast majority of cases this assistance is useful, but when considered within the context of high reliability design, the rare instances of fault introduction like this example are still frequent enough to be of concern, and should be anticipated where possible and then mitigated through defensive design techniques. Thus, if a circuit has any potential to be selected by a tool to be "improved," or if the application of Roy's Rule number 6 indicates that the circuit topology has any potential for inherent failure modes, it should be designed in a way to be bulletproof or to minimize the chance that the tool will recognize it.

Common circuit topologies, such as counters and finite state machines, are the most likely to be modified by synthesis tools. The essence of this problem is that the synthesis tool is recognizing the circuit and acting on it in an attempt to optimize it, but it is doing it at the expense of function and robustness. In order to prevent the tools from recognizing and then modifying our circuits without resorting solely upon setting switches to pre-emptively set up the tools to disable optimization features that cannot be fully predicted, the approach contained in Roy's Rule number 3 can be used, which is to design logic without the use of behavioral HDL, which leaves the actual logic design to the designer rather than relying upon the synthesis tool to do the design. This approach can work well with functions that are normally designed with behavioral HDL, such as finite state machines and complex math functions such as multipliers. However, for circuits such as this example, where the logic was already designed at the RTL level, another option is to design the circuits in ways that can minimize the chance that the tool will recognize it for what it is, or as an alternative, modify the circuit topology in a way that the circuit no longer behaves like its standard form.

Figure 6.13 is the logic diagram for a modified version of the divide by three counter that breaks the terminal count feedback path with a register, which can

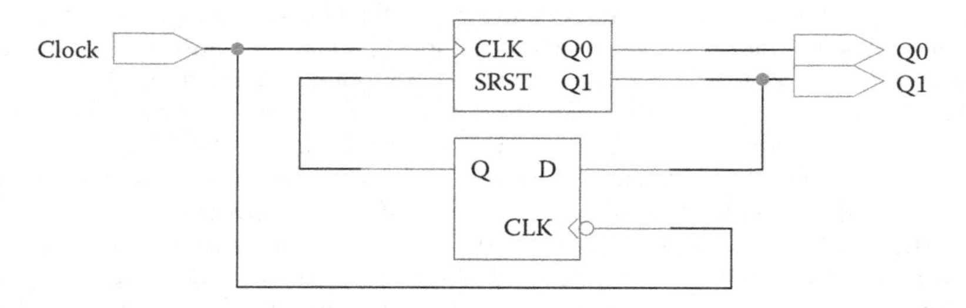

FIGURE 6.13 Modulo Counter with Broken Feedback Path

reduce the probability that a synthesis tool will recognize the circuit as being a modulo counter. The addition of the register also alters the behavior of the counter such that, while it still counts in the normal manner, it does not rely on asynchronously feeding back the terminal count.

When this circuit is processed by the synthesis tool using the same settings as the original design, the result is the netlist shown in **Figure 6.14**. It is immediately apparent that this circuit is identical to the expected synthesis result shown in **Figure 6.11** except for the addition of the delay register in the feedback path. Analysis of its operation results in the truth table shown in **Figure 6.14**, showing that the circuit's operation is also identical to the original design. Thus, with the addition of the register this counter was able to avoid the problematic feature of the synthesis tool and retain its original (and safe) functionality.

Note that some synthesis tool manufacturers have come to recognize this concern and have introduced tools with a "safe" option that bypasses the optimization features that can introduce fault modes. This is a very welcome feature for those who work with high reliability designs. However, even with this nice addition it still behooves designers to continue to practice safe and defensive design techniques.

This example illustrates the problems that can occur if designers are too willing to trust their tools. Trusting their tools is, in fact, the first impulse of most designers. One insidious reality of tool-induced failure modes is that most, if not all, of these failure modes are latent and will not be revealed through normal requirements-based verification. The designs will function as they were intended to and still meet their requirements, and so will pass all requirements-based tests. Finding these tool-induced failure modes requires additional robustness verification that

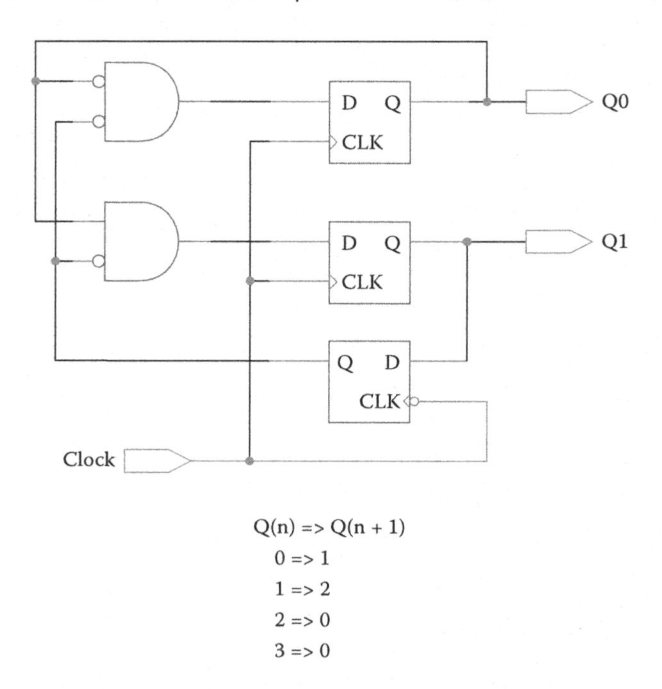

Q(n) => Q(n + 1)
0 => 1
1 => 2
2 => 0
3 => 0

FIGURE 6.14 Actual Synthesis Result for Modulo-3 Counter with Broken Feedback Path

targets the circuit types that can contain potential failure modes, such as finite state machines and counters.

When dealing with complex tools that cannot be completely understood it is important that the designer know as much as possible about the tools, but never to trust them. In addition, designers should disable as many of the advanced features as practical, and never assume that the tools are going to correctly implement the design's intended functionality. Designers should also be aware of any inferred functionality in their designs, such as unused states in counters and state machines, and create verification standards that account for all conceivable tool-induced errors.

The material presented here is not a comprehensive treatment of DATDP. DATDP is, in its purest or fullest incarnation, a lifestyle as well as a paradigm, and as such a single chapter cannot adequately express all of its aspects to their full extent. However, the introduction presented here can provide a starting point for a first step into that lifestyle, whether it is based on the information provided here or on an equivalent set of ideas that are customized for the organization or individual. So whether it is called Roy's Rules, Larry's Laws, Mark's Mandates, or any other name, DATDP should, in some form, be an integral part of any high integrity design environment.

7 Verification

Verification is performed to ensure that the requirements are met by the hardware. Verification is performed for all requirements, regardless of their origin as allocated, decomposed, or derived. The design implementation (the hardware) is evaluated against its requirements to demonstrate that the hardware performs its intended function. Since the verification is requirements based, the higher quality the requirements, the easier it will be to perform the associated verification.

Verification activities include review, analysis, and test. Level A and B hardware require independent verification. Independence can be achieved through a technical review of data by a person other than the author. A tool can also be used to achieve independence when the tool evaluates data for correctness, such as a test results checker.

A review is a qualitative assessment while an analysis is a quantitative assessment. A review is used to check a document or data against a set of criteria. Analysis assesses the hardware design or circuits for measurable or calculated performance or behavior.

Testing is the singular method by which the behavior of the actual hardware can be observed and measured.

Verification activities, methods, and transition criteria are described in the Hardware Verification Plan. **Figure 7.1** highlights the verification aspects of the hardware development life cycle.

In this chapter, "testbench" will be used as a synonym for a test procedure used in simulation because in the world of HDL (in particular VHDL) they serve a similar purpose. "Test procedure" will be used in its literal sense, i.e., when referring to procedures used for hardware test.

An effective technique is to start with a plan to organize the verification strategy. Each hardware requirement should be assessed as to whether it is suitable for review, analysis, and/or test. The optimal method for verifying each requirement is then selected and documented. For PLD requirements, the functional simulation and post-layout timing model is best suited to the analysis environment. Timing analysis can be targeted to post-layout timing models in the simulation tool, using the static timing analysis report from the layout tool or a combination thereof. An example test plan matrix is shown in **Table 7.1**.

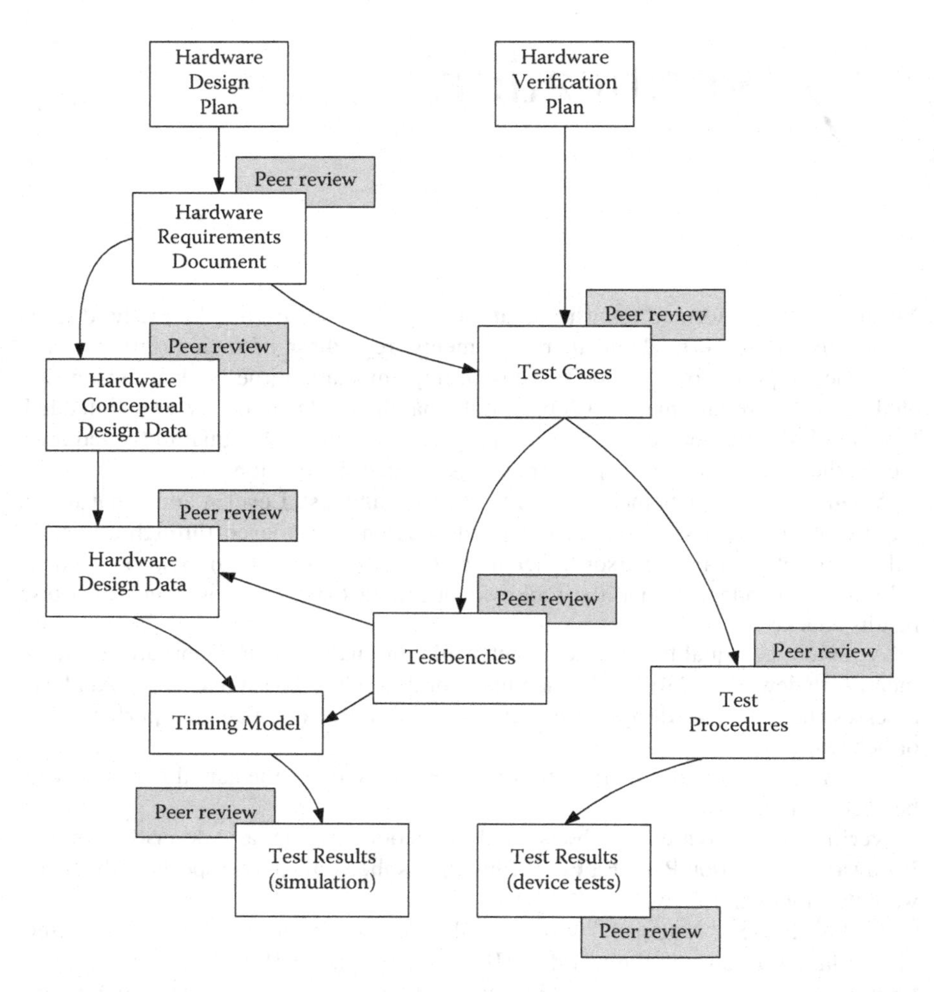

FIGURE 7.1 Verification Aspects of the Hardware Development Life Cycle

FUNCTIONAL FAILURE PATH ANALYSIS

A functional failure path analysis is performed prior to the verification process. The functional failure path analysis is used to determine which parts of the design could cause a failure that has been categorized as catastrophic (Level A) or hazardous (Level B). The circuits in the Level A and Level B functional paths are subject to additional verification methods as described in Appendix B of DO-254.

The functional failure path analysis is a top-down analysis starting from system level functional failure paths. The functional failure paths are identified for each system level function classified as catastrophic or hazardous in the functional hazard assessment. An example of a system level functional failure could be hazardously misleading data on a primary flight display or uncommanded movement of a primary flight control surface. The idea of the analysis is to identify the electronic

TABLE 7.1
Example Test Plan

Requirement ID	Review	Analysis	Simulation		In-Circuit Test	
			Functional	Post-Layout Timing	Circuit Card Tests	HW/SW Integration
Functional Element 023 (UART)						
HRD-ABC-123-001	HRD review	Dynamic timing	X	X	X	
HRD- ABC -123-002	HRD review	Dynamic timing	X	X		X
HRD- ABC -123-003	HRD review	Static timing	X			X
HRD- ABC -123-004	HRD review	Dynamic & Static timing	X	X		X
HRD- ABC -123-005	Inspection	Static timing	X		X	X

hardware in the system level functional failure path, then to decompose the electronic hardware functional failure path into the circuits constituting that path, the circuit functional failure path into the components constituting that circuit, and finally to identify the elements within the circuits.

A circuit element can be thought of as a small function that can be comprehensively analyzed and tested without further decomposition and is at the lowest level of design abstraction. In other words, the smallest functional building blocks that the design engineer assembled to create the design. For example, a circuit card designer will design a filter, not the internals of the operational amplifier used in a filter. For a circuit card, elements are circuit functions with well-bounded and well-characterized functionality that can be tested. Examples of circuit elements on a circuit card would include amplifiers, comparators, filters, analog to digital or digital to analog convertors, or discrete logic.

The size of an element can also be defined by the capabilities of the test and verification tools. For a PLD, a working definition of an element would be logic functions with 12 or fewer inputs (to ensure compatibility with code coverage tool operation). Examples of PLD elements would include RTL-level building-block functions such as counters, decoders, comparators, multiplexers, registers, or finite state machines.

An example of the decomposition of a functional failure path is shown in the following figures. **Figure 7.2** shows the system level functional failure path for the loss of motion hazard in a primary flight control system. This example identifies the contribution from the sensors on the input to the functional failure path.

Figure 7.3 shows the decomposition from the sensors functional failure path to the functional elements in the hardware on the circuit card.

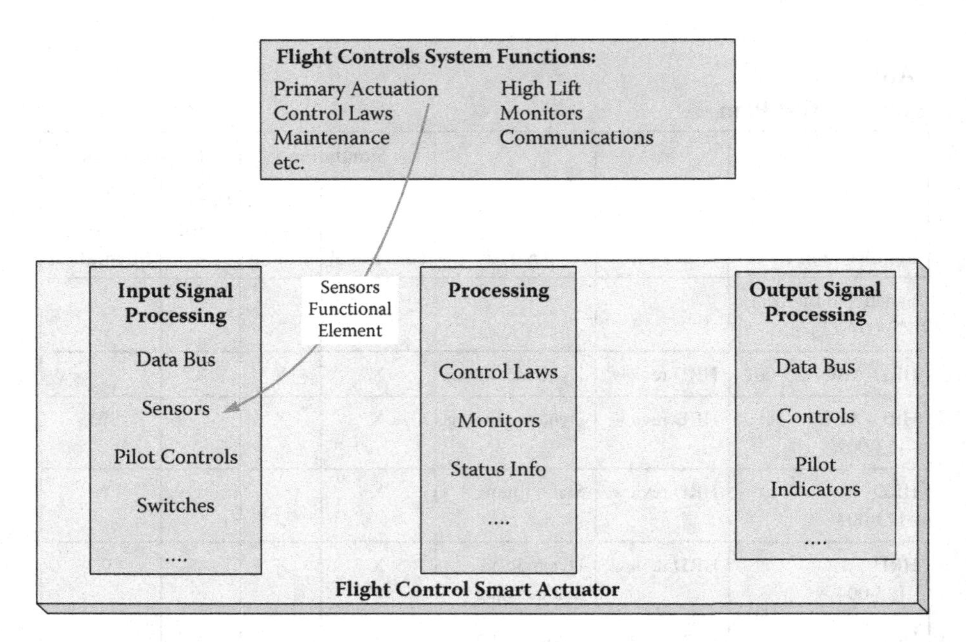

FIGURE 7.2 **(See Color Insert.)** Functional Failure Path for Loss of Motion

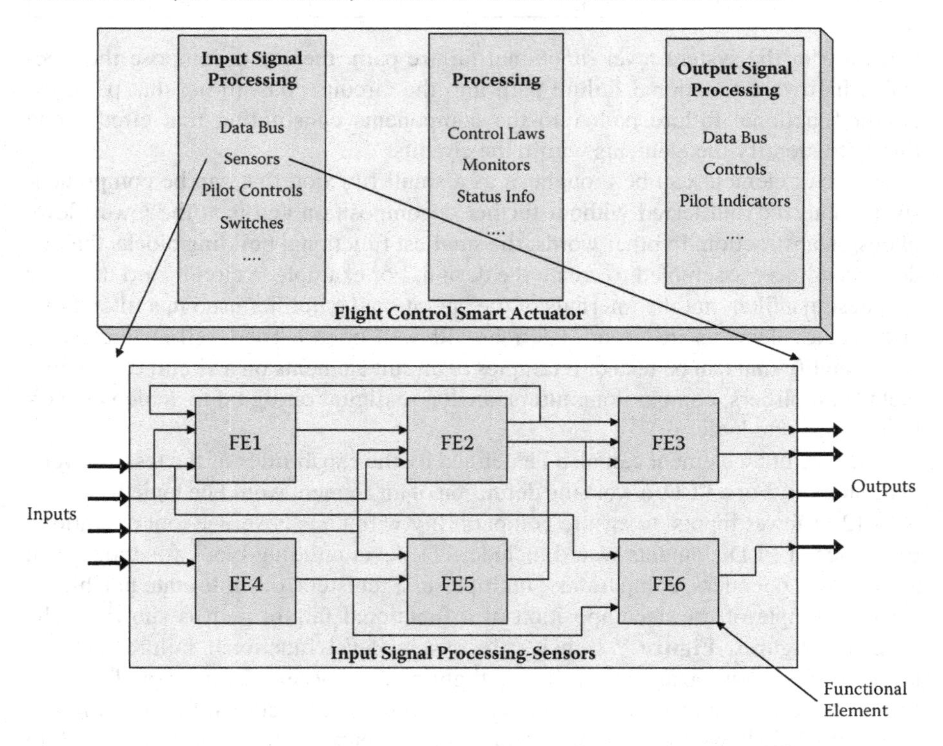

FIGURE 7.3 **(See Color Insert.)** Functional Elements within the Functional Failure Path

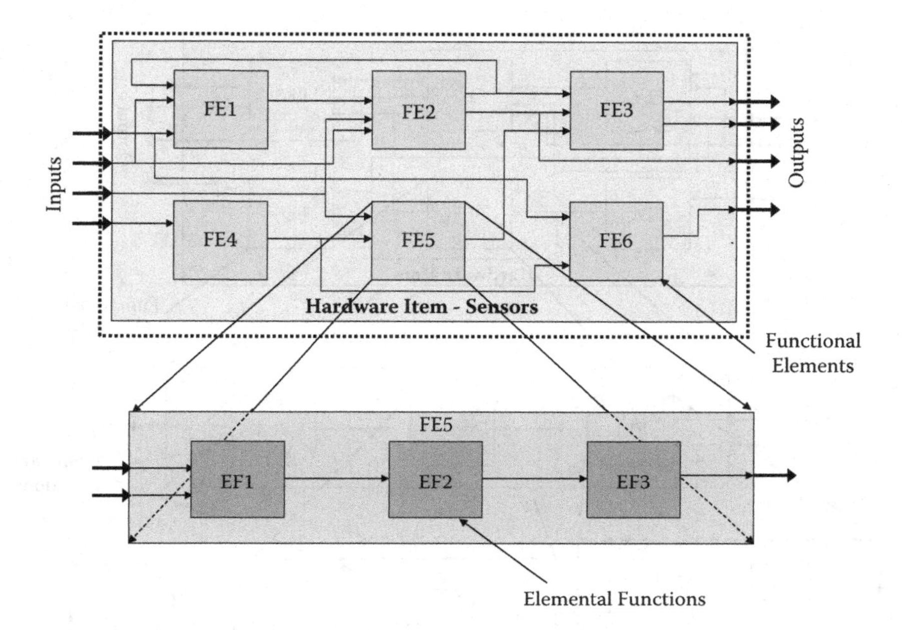

FIGURE 7.4 **(See Color Insert.)** Elements That Implement a Function

Figure 7.4 shows the decomposition of one of the functional elements on the circuit card to the elemental functions or circuits that implement the function.

And finally, **Figure 7.5** shows the circuits for each elemental function.

Another example is provided to show how a PLD works into the decomposition of functional failure paths. This is a fictitious system for illustration purposes only. This example has two functional failure paths: one for the control actuator output and the second for the position feedback and monitoring. **Figure 7.6** shows the block diagram of the system.

The decomposition starts at the top and works down to the elemental functions:

- System level to hardware level
- Hardware level to circuit level
- Circuit level to component level
- Component level to elemental level

Figure 7.7 shows the functional failure paths (control and feedback), the electronic hardware in each FFP, and the functional elements in each FFP as indicated by the dashed boxes.

The next diagram, **Figure 7.8** shows the functional elements in each functional failure path, the elemental functions in each functional element, and the components in each elemental function.

Figure 7.9 then shows the elements within each elemental function. In this example, some of the circuit functional elements have one circuit element while others have more than one. The controller for the analog to digital convertor was chosen to

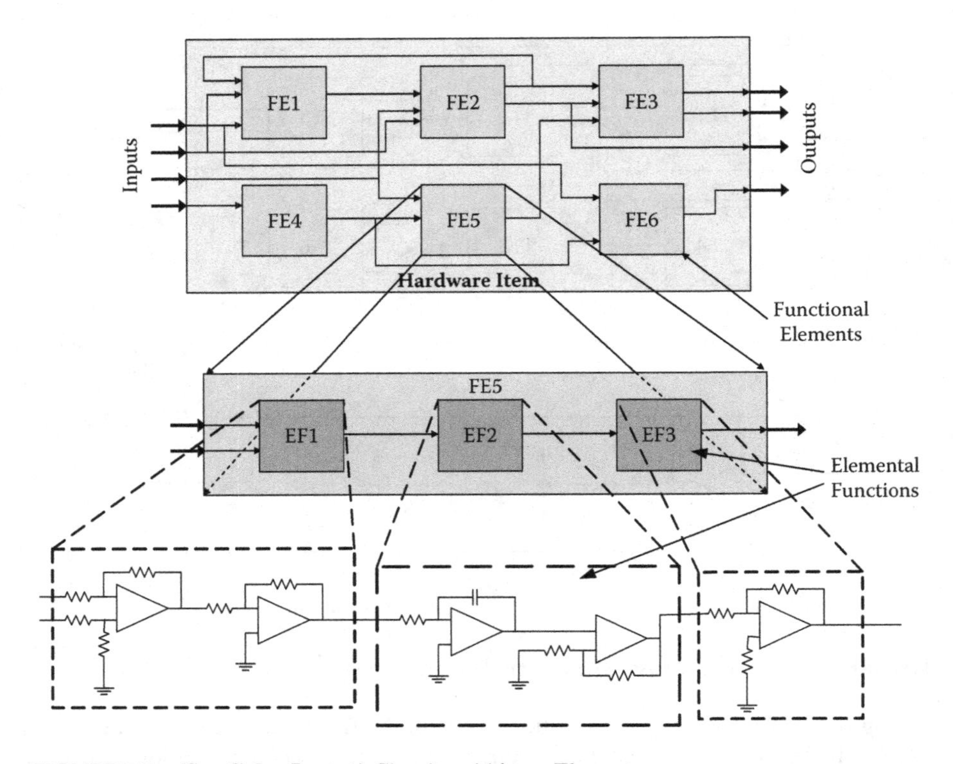

FIGURE 7.5 **(See Color Insert.)** Circuits within an Element

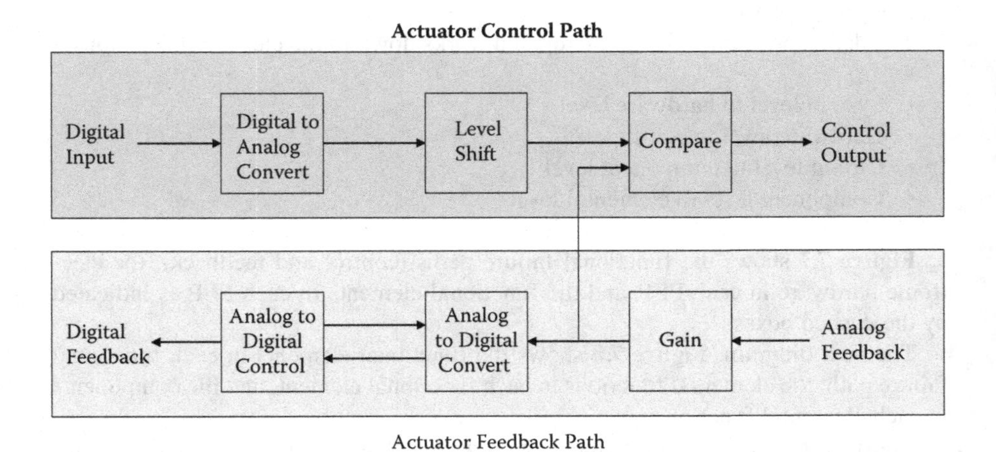

FIGURE 7.6 Flight Control with Two Functional Failure Paths

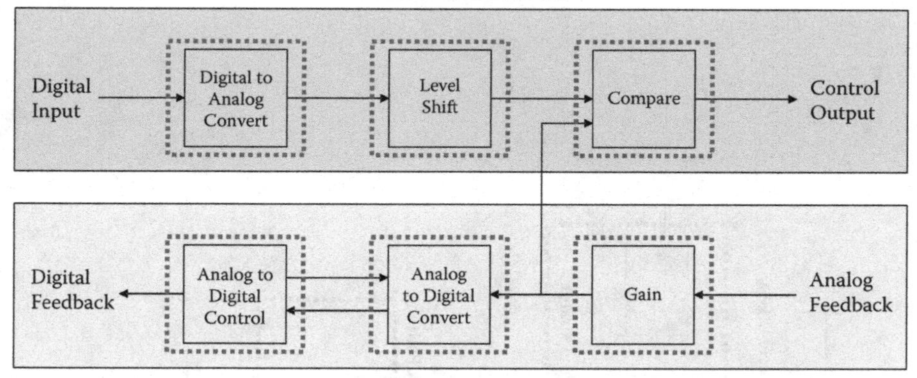

FIGURE 7.7 Functional Elements within Each Functional Failure Path

be a PLD. Since the PLD is entirely in the actuator feedback functional failure path, the design assurance level of the PLD will be the same as the design assurance level of the actuator feedback functional failure path. In this case it is Level A. The PLD is this example also shows multiple elements within the device.

When a PLD has functions that are in multiple functional failure paths, the design assurance level for the PLD is the highest level associated with the respective functional failure paths. The reason for this is that it is not possible to demonstrate that the gates, clocks, and power within the PLD are completely isolated. Even if it was possible to show the separation between Level A, B, C, and D functions within one PLD, they still share the same physical packaging and connection to power and ground.

Figure 7.10 shows the same concept for a PLD. The functional failure comes from an output with erroneous behavior. Following the suggestions in this book on writing requirements, the functional failure path will traverse one or more outputs in one or more functional elements within the PLD.

Once the functional failure path analysis is performed, an additional verification method is selected for Level A and Level B circuits or elements.

APPENDIX B ADDITIONAL VERIFICATION

Appendix B of DO-254 identifies several methods of advanced verification that can be applied to Level A and Level B hardware. At least one of these methods needs to be selected and applied. DO-254 does permit other methods to be proposed for additional verification, subject to certification authority agreement. The advanced verification methods identified in DO-254 include elemental analysis, safety-specific analysis, and formal methods. This book will concentrate on elemental analysis since it is by far the most commonly used and is supported by commercially available tools.

Elemental analysis provides metrics on how much of the design (elemental functions) was covered through the requirements-based verification of the associated

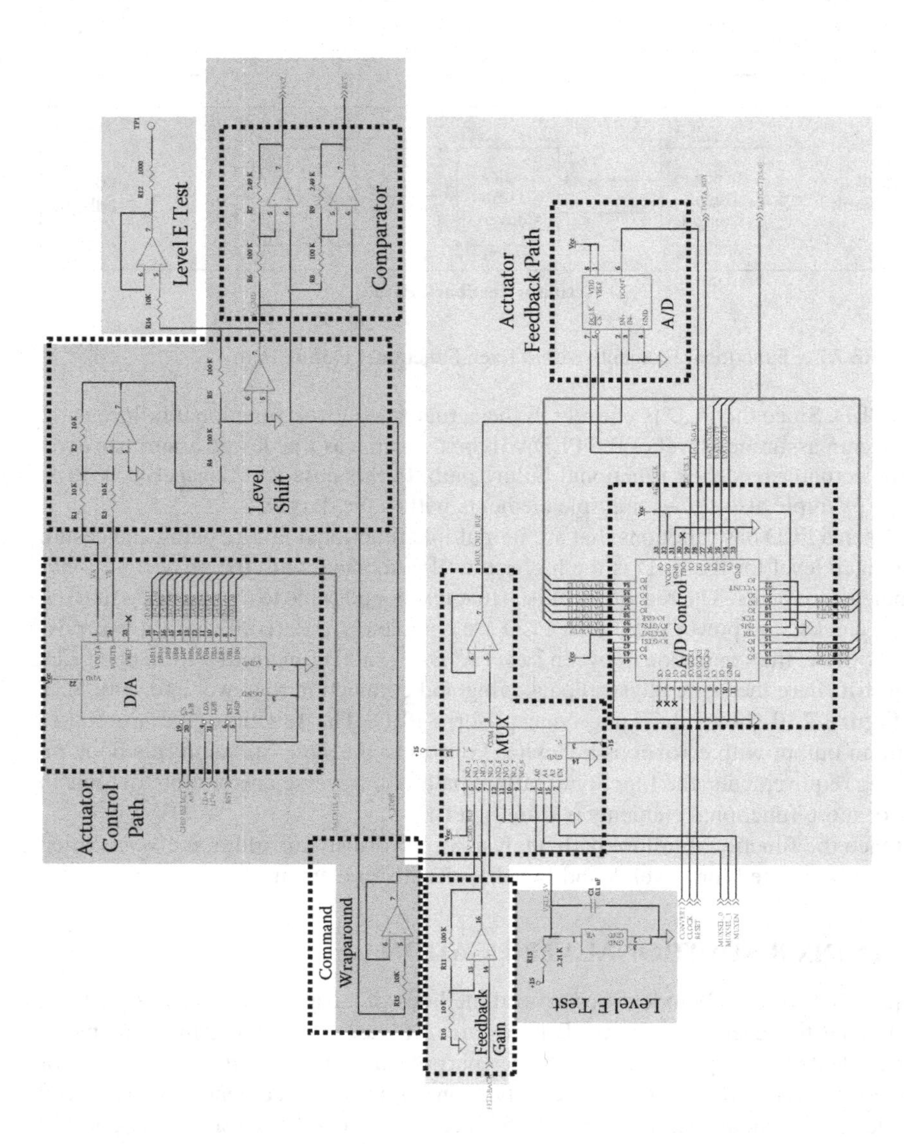

FIGURE 7.8 **(See Color Insert.)** Functional Elements Overlaid on the Design

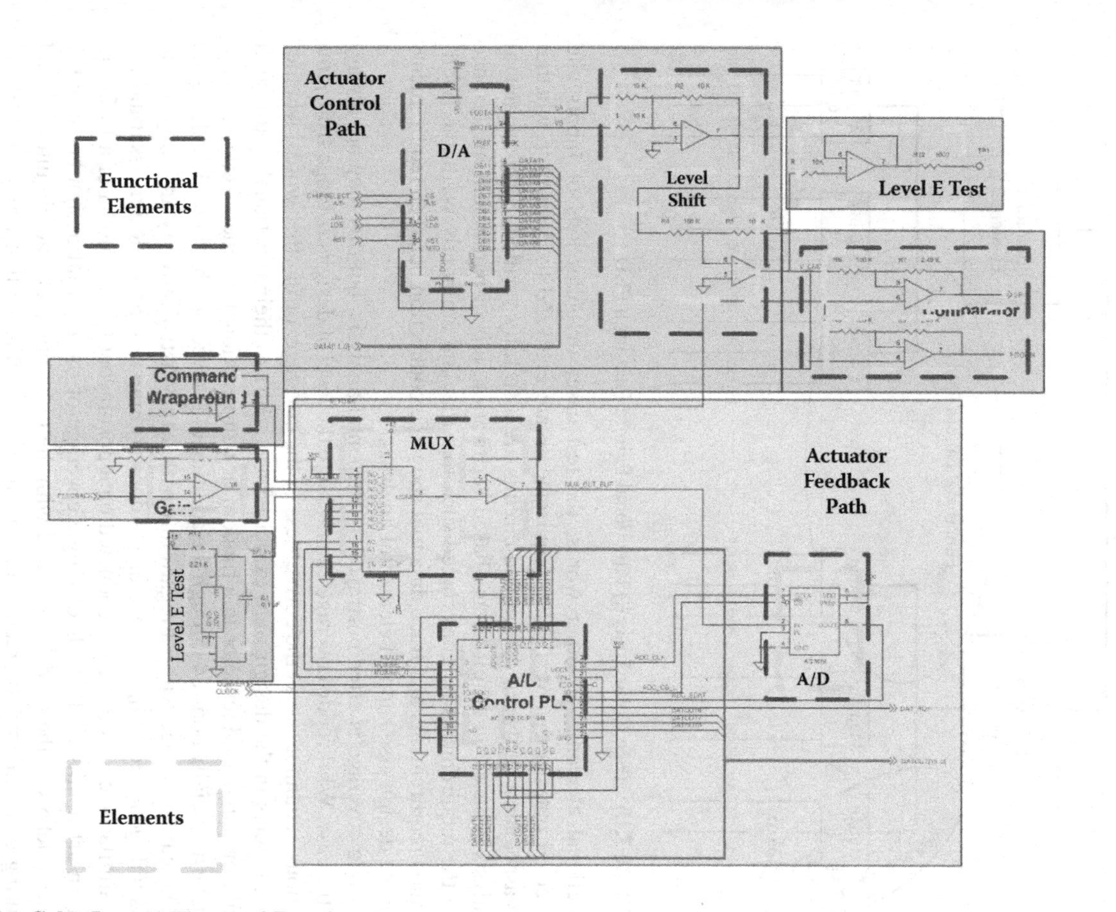

FIGURE 7.9 **(See Color Insert.)** Elemental Functions

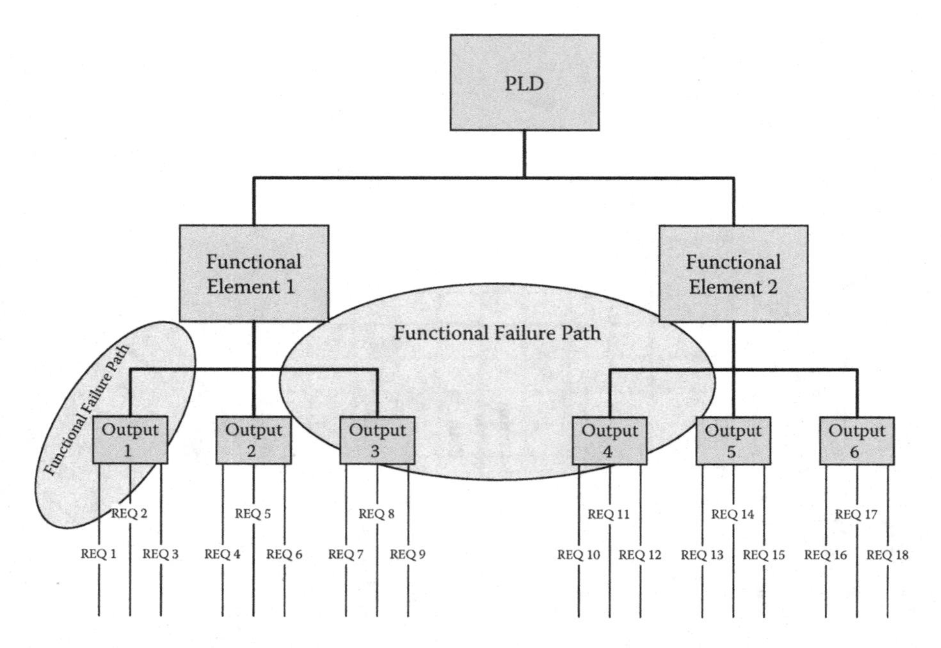

FIGURE 7.10 **(See Color Insert.)** Functional Failure Path for a PLD

functional elements. In other words, the test cases or collections of inputs and expected outputs that verify the requirements are analyzed to determine whether the associated hardware design and circuits are also fully verified. Since elemental analysis is based on requirements-based verification, it is of paramount importance that the requirements be constructed such that they are eminently verifiable. Well-formed requirements and trace data that show the connection between requirements and the design also support elemental analysis.

Code coverage is often portrayed as elemental analysis, but this may not always be the case. While code coverage can measure the level of design coverage from simulations, it cannot assess the correctness of the coverage, or in other words, whether the elements in the design were exercised according to their functionality (through requirements-based tests) or in a random or non-functional manner. Code coverage can be artificially boosted through the addition of test vectors that are solely designed to increase coverage without regard to whether they actually verify anything; while this will generate complete coverage, it does nothing to support design assurance.

Similarly, code coverage conducted on a post-layout timing model of a PLD can return artificially high levels of toggle coverage of the internal design because the timing delays in the model combined with the often complex and circuitous signal routing can combine to create race conditions and glitches that imitate legitimate logic combinations. There is also little correlation between the elements (lines of code) in the design (the HDL source code) and the nets and nodes in the post-layout timing model, so code coverage of the inside of the post-layout timing model has little to no real utility. In addition, DO-254 states that elemental analysis should be conducted at the level at which the designer expressed the design (in this case the

HDL elements, which are nominally lines of code), whereas synthesized and routed netlists express the design at a significantly lower level of expression than the actual design. Thus code coverage is only useful when measured during functional simulations (simulations of the HDL source code as opposed to the post-layout timing model or post-synthesis netlist), or when toggle coverage is measured for inputs and outputs during simulations on the post-layout timing model. Measuring toggle coverage during post-layout simulations may be unnecessary if a review of the test cases indicates that all inputs and outputs were toggled appropriately during the normal course of post-layout simulations; however, since code coverage metrics are measured in the background during simulations, they can normally be gathered during simulations with less trouble and more accuracy.

For analog circuits and circuits not implemented in a PLD, no amount of code coverage will reveal how much of the design was covered by the requirements-based tests. While not frequently used these days, PLDs specified with schematic capture would need an analysis of the design as specified by the designer, not code coverage. Code that is written as behavioral code or translated from a higher level language such as SystemC can also be problematic since the designer specifies functionality, not the design itself. Code coverage for elemental analysis for a PLD works best when the requirements are specified functionally and written at the pin level and the associated design is expressed at the register transfer level (RTL). An RTL design will have very close alignment with pin level requirements.

Using the definition for elemental analysis in DO-254 and a PLD that has functional requirements written at the pin level and valid trace data to a RTL level definitions, the analysis is complete when the requirements-based verification is complete. How is this possible? If the requirements are comprehensively verified, all the outputs will have been covered and all the inputs will have been covered. Since the design is closely related to the requirements and there are no extraneous elements in the design—as shown through top-down and bottom-up traceability—comprehensive requirements-based verification can be shown to also be comprehensive verification coverage of the design. This alignment is a natural byproduct of writing requirements from output back to input with a template that covers power on conditions, response to reset conditions, and how the output(s) asserts and deasserts in response to the associated input(s). **Figure 7.11** shows how requirements and the related design align.

The test cases to verify the requirements along with the trace data that shows the connection between requirements and design will demonstrate that the design is fully covered through the requirements-based verification. Code coverage should still be performed to satisfy FAA Order 8110.105. Alternatively, code coverage for elemental analysis can be proposed when there is suitable alignment and traceability between the requirements, the design, and the test cases.

Elemental analysis, in particular code coverage, can reveal elements that were not fully exercised by the requirements-based test cases. When a coverage deficiency is revealed, the first impulse may be to analyze the deficiency to identify the missing input combinations and then add those combinations to the test case. While this approach may eliminate the deficiency and increase coverage, care must be exercised to ensure that this is actually the correct response.

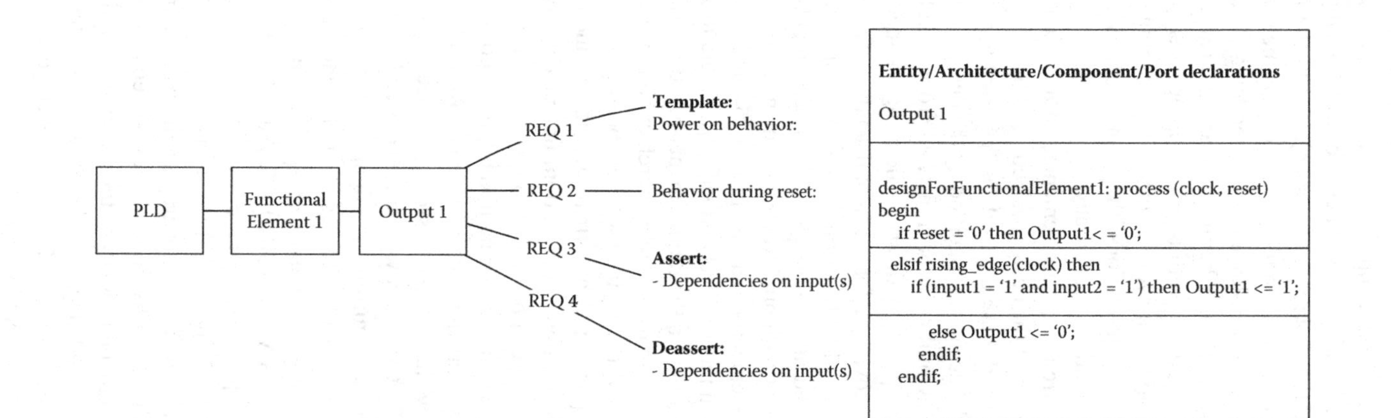

FIGURE 7.11 Relationship between Functional Elements and the Elemental Functions in a HDL Design

The goal of elemental analysis is not to get full coverage of the design. In fact, it can be argued that elemental analysis has no goal; it just fulfills the function of measuring the verification coverage of the design—full coverage of the design is the goal of the requirements-based verification. Therefore, when elemental analysis indicates less than full coverage of the design, it should not be interpreted as some kind of failure.

Coverage deficiencies can be caused by inadequate test cases, design errors, unused design features, extraneous functionality, defensive design practices, or requirements issues such as excess, missing, erroneous, or inadequate requirements. Each deficiency in coverage should be analyzed to identify the cause of the deficiency, and only then should the deficiency be addressed.

The first impulse for correcting a deficiency—adding input combinations solely to eliminate deficiencies—will not directly address any of the possible causes, so jumping to that solution without an analysis is not a sound approach. If additional input combinations are desired, they should be added methodically and deliberately through an analysis of the applicable test case to identify how the applicable requirement was not fully verified. Only when the analysis reveals a deficiency in the test case (i.e., an aspect of the requirement that was not verified by the test case) should additional inputs be added, and even then the additional inputs should be based on verifying the unaddressed aspect of the requirement, not the inputs that would eliminate the deficiency. If an analysis indicates that the test case completely verified all aspects of the requirements, additional analyses should be conducted to determine whether the coverage deficiency is a symptom of a more serious issue. On the other hand, if an analysis shows that the deficiency was just a matter of selecting a different data pattern on a data bus (and where the exact data pattern had no functional significance), then it would be appropriate to change or add the input data to eliminate the deficiency.

INDEPENDENCE

DO-254 requires independent verification of all Level A and B functions. Independence is achieved when the verification is performed by a person, tool, or process that is different than the designer or author of the data. Test cases and analysis may be created by the hardware designer as long as the subsequent review of the test cases and analysis is performed by someone other than the designer. Since the designer has intimate knowledge of the design of the hardware, there is a potential that their verification efforts are biased toward checking that the hardware operates as designed. In other words, a designer can induce a common mode error by verifying that the hardware is verified against its intended design rather than its intended functionality. Independence is used to ensure that hardware is verified for intended function in accordance with the requirements.

Independent reviewers can work in the same team. DO-254 does not specify that designers have to be in the design team and reviewers in the verification team. Tools can also be used to achieve independence. For example, a suitably assessed or qualified simulation tool can use automatic results checking rather than requiring a person to check each and every simulation result.

The above discussion of independence, as well as the definition used in DO-254, relies solely on functional or role-based independence, i.e., independence that is defined by whether a reviewer is independent of the material being reviewed. There is also a more subtle component of independence that is not addressed in DO-254 and may be overlooked due to its subtlety, but can undermine independence if it is not recognized and mitigated. That component is cultural independence. In the context of this discussion, "cultural" refers to the prevailing engineering culture of the organization or even of the corporation. The means by which an engineering culture can undermine independence is not always apparent and can be difficult to recognize, let alone mitigate, but its effects can be real and detrimental to the integrity of design assurance.

An organization's engineering culture can (and often does) influence or even dominate peer reviews, or for that matter, virtually all aspects of engineering. In most cases, especially when the culture stresses integrity and thoroughness, the effect of the culture is a positive one. In some cases, however, a strong engineering culture can make it difficult to ensure that peer reviews are actually independent despite compliance to the independence criteria in DO-254. If an engineering department has institutionalized practices and standards that are less than adequate for the required design assurance, a reviewer that comes from that same environment may not be able to recognize deficiencies even though they conscientiously look for them. Or in other words, if a designer creates an unsafe design feature in the hardware because the engineering standards and culture incorrectly endorse that unsafe feature as being acceptable, and the reviewer is a product of that same culture and standards, then despite being independent of the design the reviewer will incorrectly interpret the unsafe design feature as being acceptable.

Another (and arguably more common) example of this phenomenon can occur when writing and reviewing requirements. If the prevailing engineering culture has adopted poor requirements practices and has de-emphasized the generation and use of high quality requirements to the point where no one in the organization can even recognize a poorly written requirement, and both the requirements author and reviewer are products of that culture, then even an independent requirements review will only confirm that the poorly written requirements were "correctly" written in compliance with the prevailing standards. This effect can occur even if the author and reviewer are from different departments (such as design versus verification departments).

Thus if the cultural aspect of independence is not accounted for, there is a real possibility that effective independence cannot be achieved even when complying with the DO-254 definition of independence. Mitigating this phenomenon may be difficult, but if the selection of reviewers takes this cultural influence into consideration, or the review checklist uses review criteria that are detailed and reflect known high integrity concepts, its effect can be reduced if not actually eliminated.

REVIEW

Reviews are performed on all DO-254 life cycle data. Level A and B hardware require an independent review in accordance with Appendix A of DO-254. Reviews are typically documented with a checklist that lists the reviewer, item reviewed, and the criteria used for the review. The review checklist should be referenced from the Hardware Verification Plan or Hardware Verification Standards and stored in the configuration management system in accordance with the Hardware Configuration Management Plan. The review procedures or checklists should be HC1 controlled for Level A and B hardware. The completed checklist is stored as HC2 controlled data in the configuration management system in accordance with the Hardware Configuration Management Plan.

Reviews of hardware documents and data are accomplished by initially performing a full review. Once the full review of the data or document has been performed, incremental reviews can be used for any subsequent changes. That is, the review criteria only need to be applied to changes made to the data or document.

When performing the review, all checklist questions should be answered. Any comments should be noted on the checklist and discussed with the document author or technical authority. Each comment should document an agreed corrective action. Once the agreed comments and actions are resolved, the reviewer can verify that the document has been updated appropriately.

Checklists present a series of questions that are selected to ensure that the hardware design life cycle data has the following characteristics:

- Unambiguous—the information and/or data is documented in such a manner that it only permits a single interpretation.
- Complete—the information and/or data includes all the necessary and relevant descriptive material that is consistent with the standards. The figures are clearly labeled, all terms are defined, and units of measure are specified.
- Verifiable—the information and/or data can be reviewed, analyzed, or tested for correctness.
- Consistent—the information and/or data has no conflicts within the document or with other documents.
- Modifiable—the information and/or data is structured in such a way that changes, updates, or modifications can be completely, consistently, and correctly made within the existing structure.
- Traceable—the origin or derivation of the information and/or data can be demonstrated.

Peer review checklists should also include criteria to check all aspects of hardware standards for reviews of requirements, design, verification, and validation data.

Table 7.2 lists the reviews to be performed and each associated review checklist. If the data item is not required for the design assurance level, then the associated review is not needed.

Reviews of hardware management plans and standards are accomplished by a full document review. Document reviews are also used for the review of the Hardware Configuration Index, Hardware Environment Configuration Index, and the Hardware Accomplishment Summary. Incremental reviews can be used for subsequent changes.

Reviews of requirements should apply the checklist questions or criteria to each and every requirement. For convenience, the reviewer can create a spreadsheet to log

TABLE 7.2
Life Cycle Data Item and Associated Review Checklist

Life Cycle Artifact Reviewed	Review Checklist
Plan for Hardware Aspects of Certification	Plan for Hardware Aspects of Certification Review Checklist
Hardware Design Plan	Hardware Design Plan Review Checklist
Hardware Verification Plan	Hardware Verification Plan Review Checklist
Hardware Process Assurance Plan	Hardware Process Assurance Plan Review Checklist
Hardware Configuration Management Plan	Hardware Configuration Management Plan Review Checklist
Hardware Requirements Standards	Hardware Requirements Standards Review Checklist
Hardware Design/Code Standards	Hardware Design/Code Standards Review Checklist
Hardware Requirements Document	Hardware Requirements Document Review Checklist
Hardware Design Data	Hardware Design Data Review Checklist
Test Cases	Test Case Review Checklist
Test Procedure/Testbench	Test Procedure/Testbench Review Checklist
Test Results	Test Results Review Checklist
Hardware Verification Report	Hardware Verification Report Review Checklist
Hardware Configuration Index	Hardware Configuration Index Review Checklist
Hardware Environment Configuration Index	Hardware Environment Configuration Index Review Checklist
Hardware Accomplishment Summary	Hardware Accomplishment Summary Review Checklist

the checklist response on a requirement-by-requirement basis. Reviews of the design apply the checklist questions to each and every schematic (or file for HDL-based designs). The reviewer can create multiple tabs on a spreadsheet to log the checklist response on a file-by-file basis.

Reviews of test cases apply the checklist questions to each and every test case within a test case file. The reviewer can create a spreadsheet for each test case file to log the checklist response on a test case basis.

Reviews of the test procedures and testbenches apply the checklist questions to each and every test procedure or testbench file. The reviewer can create a spreadsheet to log the checklist response on a file-by-file basis.

Reviews of the test results or simulation results apply the checklist questions to each and every test or simulation results file. The reviewer can create a spreadsheet to record the checklist response on a file-by-file basis.

Reviews should clearly list the findings, discrepancies, and any items that do not satisfy the review criteria. The author should make the necessary updates to resolve the review comments, or provide a response to clarify any misunderstandings on the part of the reviewer. Once the document or data has been updated, the reviewer should check the updates and close out the review comments. It is helpful for the review checklist to specify the configuration identifier and version of the document initially reviewed and the final corrected version of the document.

ANALYSIS

As stated above, analysis is used for quantitative assessment of the hardware design. Circuit designs typically use thermal analysis to ensure that the design works within specified tolerances at extreme high and low temperatures. Circuits are also analyzed to ensure that the design works as specified when component tolerances vary. Reliability analysis is used to determine whether the actual implementation of the design meets reliability requirements. Designs can be compared to previously approved designs with a similarity analysis.

PLD designs use an analysis typically referred to as simulation to verify that the design meets its functional requirements. Simulations are computing resource intensive, so for best performance it is recommended that simulations be run on the most capable computing platforms available. In general, Linux-hosted systems can provide the best performance for execution of simulations. Workstations or servers running the simulation should be configured with adequate random access memory, a high speed or solid state disk, and high speed Ethernet controller. Faster workstation or server clock speed and multicore processors will also accelerate simulation performance. Running simulations on laptop computers will reduce performance and can interfere with staff productivity if the laptop is also used for office applications such as electronic mail and document editing. Tool vendors can provide information on the optimal environment for hosting simulation tools.

Once the test cases are selected for comprehensive coverage of the requirements, the testbenches are created to execute the simulation. These testbenches can be run in multiple scenarios or configurations to collect various types of data and metrics. **Figure 7.12** shows the scenarios used for simulation of a PLD design.

Functional simulations are conducted on the HDL and apply the input stimuli defined in the test cases and implemented in the testbenches. These simulations run quicker than timing simulations and allow debug and design tuning. Functional simulations can also use code coverage tools to collect coverage metrics. These coverage metrics can be used to support the elemental analysis.

Post-layout simulations are conducted on the post-layout (sometimes called a post-route) timing model, which is the netlist output of the place and route tool that has been annotated with the representative timing delays of the target PLD device. This timing model is the most realistic representation possible of the programmed PLD and allows simulation results to be conducted on a model that is as faithful as possible to the programmed device. Post-layout simulations with timing data allow for worst-case timing analysis, typical timing analysis, and best-case timing

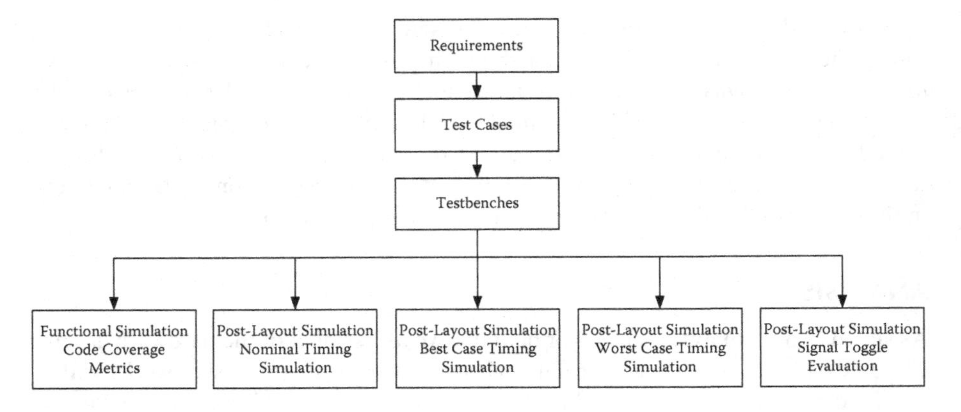

FIGURE 7.12 PLD Simulation

analysis. These simulations are used to demonstrate that the device requirements are met under all timing variations. Typical timing simulations are useful when comparisons with device tests are needed. Post-layout simulation can also verify toggle coverage—this shows whether all inputs and outputs have transitioned high and low during the testbench execution. Toggle coverage verifies that device signals are present and used, and that static inputs and unused outputs do not toggle.

Hardware testing for DO-254 requirements-based tests are typically performed at room temperature. The hardware is built with the actual circuit components, but it would not be practical to select components with the most extreme tolerances. Further, testing would require multiple runs with components selected for best case timing or tolerance and again for worst case timing or tolerance. Post-layout simulations allow efficient evaluation of PLD behavior at minimum and maximum propagation delays. This allows hardware testing to use standard rather than screened or selected components. Timing simulations also provide evidence that the PLD will meet its requirements and provide the correct signals to peripheral devices for any combination of temperature, clock, or voltage variations, or variations of the PLD or external components that the PLD interfaces with.

An analysis performed for verification credit should state the procedure used, the configuration identification of the data analyzed, the person who performed the review, the results of the analysis, any actions needed to correct the design or the data, and the conclusion or summary of the results.

When simulation is used for PLD analysis, all raw data files from the simulation tool should be collected. Waveforms should be exported to a file and, at a minimum, archived for the project. Test cases can be indexed and the index included in the waveform to allow quick lookup within the waveform file. A test case counter is an effective way to index through a waveform file. Some engineers use "do" files to quickly index timelines in waveform files. The "do" files should be listed in the verification report and archived with the verification data.

If the simulation waveforms will not be manually analyzed to compare actual timing and logic levels to the expected results in the test cases, or in other words if the testbenches will evaluate the data to derive the pass/fail results, the testbenches should be written to output text log files that document the waveform data. These log files should be verbose and include the following:

- The date, time, name of tester, and configuration identifier of the testbench.
- The configuration identifier of the design under test (being simulated).
- Every stimulus applied to the design under test, along with the timestamp for the stimuli.
- Every output from the design under test, along with the timestamps for the outputs.
- For each pass/fail evaluation, the expected result, actual result, pass/fail evaluation, and timestamp.

Simulation log files that simply state PASS or FAIL (whether for each expected result or for the entire test case) are not sufficient evidence for compliance to DO-254 verification objectives.

In addition, the testbench code that evaluates the data and generates the pass/fail result should be designed to be easily understood and characterized such that a code review will determine to a high level of confidence that the code will correctly evaluate the results. Obviously the level of confidence in the testbench code should be commensurate with the design assurance level of the design. There is no official guidance on how such testbench code should be written, nor for how it should work, but the wise approach is to make it as simple and as easily understood as possible to minimize the probability of errors and thus of concerns from the certification authorities.

The summary of the analysis will summarize the verification highlights and results. The conclusion of the analysis will provide a conclusion based on the data. A statement as to whether the PLD met its operational requirements for the specified timing, voltage, clock, and temperature should be included.

DO-254 Section 6.2.2 also mentions that analysis should be performed to assess whether the verification is complete. The verification coverage analysis determines whether all requirements have been verified by a review, analysis, or test. The verification can be performed at the most suitable hierarchy of the design. PLDs can be electrically tested with circuit card level tests, in conjunction with software tests, or in the course of system or higher level tests. Note that the higher level tests still require that the PLD inputs and outputs be controlled, observed, and recorded. The verification coverage analysis also determines whether the review, analysis, or test was appropriate for the requirement. System level tests conducted in a closed box manner would not be suitable for instrumenting a PLD for requirements-based testing. Output signals from a PLD that go directly to a box level output, and which allow the PLD timing to be observed, could conceivably be used for the corresponding PLD requirements. An examination of the circuits between the PLD and the box output would be needed to determine how those circuits affect the signal level and timing. Finally, the verification coverage analysis determines whether the results are correct. Any differences between actual results and expected results are captured and explained. Particular attention should be paid to verification results for requirements designated as safety related, especially if the verification results are not exactly as expected.

TEST

Testing is the verification of the actual hardware to show that the implementation meets the requirements. Testing is typically performed in-circuit with production or production equivalent hardware. Production equivalent means that the circuits and card layouts are the same as those that will be used in the aircraft. Any differences should be assessed for impact. Conformal coating on the circuit cards would interfere with electrical testing and is one of the typical differences between production and production equivalent electronic hardware. Test headers or sockets for devices to allow access to pins are also typical changes to the board under test. It is essential that PLDs be the same device type and programmed with the same content and same programming methods as those used in the manufacturing environment.

Testing can be performed at the system level, circuit card level, circuit level, or component level. DO-254 recommends that testing be performed at the highest level of integration possible within the system. Testing in a more integrated manner allows for easier evaluation of interfaces and detection of unintended side effects. Hardware testing for a processor card can take advantage of software tests performed for DO-178C credit. Hardware testing can also use existing production, environmental, and acceptance tests as long as the correlation and traceability to requirements is possible. Regardless of the type of test bed used, the test should instrument the circuit and design to allow the inputs and outputs to be recorded during the course of a test. Some instrumentation of circuit cards and PLDs is not possible during closed box system tests. Closed box tests would not be a suitable test environment for many requirements-based tests necessary for DO-254 compliance.

Once the test cases are selected for comprehensive coverage of the requirements, the test procedures are created to execute the test. These test procedures can be run on circuit cards, at the perimeter of a PLD mounted on a circuit card, or on a standalone device tester for a PLD. Test results are collected with oscilloscopes, voltage meters, logic analyzers, and other test equipment. **Figure 7.13** shows the scenarios used for testing a PLD design.

TEST CASE SELECTION CRITERIA

While DO-254 does not explicitly require test cases as a separate data item, experience has shown that there are advantages to creating test cases that are separate from test procedures. The test cases are the set of inputs applied to the hardware or design with a corresponding set of predicted expected results and pass/fail criteria. The test cases are used to select the set of inputs that will be applied in a hardware test or a simulation.

With requirements that express pin level behavior, the process for verification can be optimized and more efficient. The first opportunity for improving processes is to

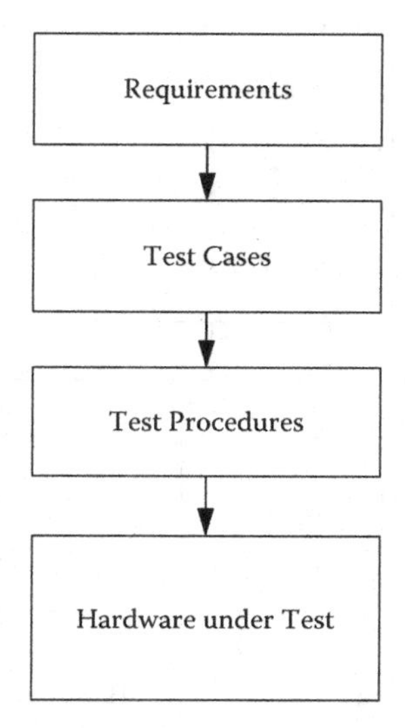

FIGURE 7.13 Hardware Test

create the test cases against the requirements as soon as the requirements are mature and have been reviewed. This allows the verification activities for testing and simulation to start even before the design is created.

The second opportunity for improving processes is to base the test cases entirely on the requirements, and where possible, write them to be independent of any particular verification method (in particular simulation and hardware test). This will allow the same test cases to be used for hardware test, simulation, or where appropriate, both. For hardware test, the test procedures are written from the test cases to implement and apply the input stimuli in a physical hardware test environment. For simulation, testbenches are written from the test cases to implement and apply the input stimuli in a simulated or virtual environment.

A third area for efficiency with test cases is to format and organize the test cases to allow the reviewer to more quickly assess whether the requirement(s) have been comprehensively verified. Test cases can be formatted in tables with the inputs and outputs in columns and the set of input values and expected results across each row. The use of partitioning to separate individual test cases, allowing each test case to be more easily identified and correlated to individual requirements or even individual parameters within requirements, will aid in assessing and reviewing test cases and test procedures. Attaching some means of identification to steps or operations within test cases (such as step numbers) will also allow for easier tracking of the testing process and permit more accurate references to individual features of the test case.

Separating test cases from the procedures and benches also allows the analytical aspects of the effort to be concentrated on the test case design. Writing the test procedures and testbenches becomes a more rote task of translating the input stimulus to the respective environment.

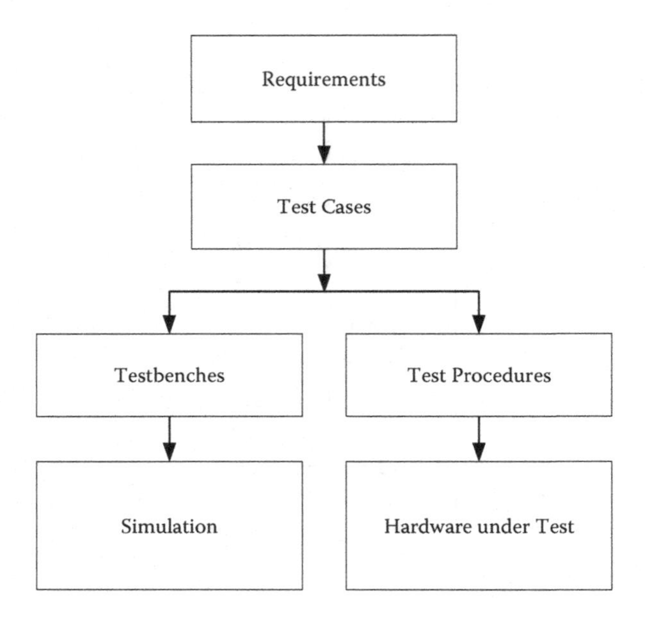

FIGURE 7.14 Common Test Cases

Figure 7.14 depicts the test case paradigm described above.

With requirements formulated as described in the requirements chapter, another optimization occurs: writing requirements that state an output in response to an input ensures that each requirement is verifiable—an input can be applied and an output can be observed. The inputs in the statement of the requirement become the inputs for the test case. The outputs in the statement of the requirement become the outputs for the test case. The transfer function or behavior expressed in the requirement allow the test case author to readily predict the output and expected results in the presence of the inputs. The structure and correlation of requirements and test cases is shown in **Figure 7.15**.

When translating test cases to simulation testbenches, the test case expected result can be evaluated against the simulation waveform through the use of self-checking code. The simulation log file should also state the expected result, the actual measured result, timestamp, and the pass/fail of the comparison, as described previously. For hardware testing, the event in the test case from which the timing for the expected result is measured, or the expected result itself if appropriate, can be translated into the trigger conditions for a logic analyzer or oscilloscope. Also note that simulation waveforms and waveforms from hardware test could be compared side by side with the expected results to make the results review more efficient.

While inputs in a simulation can be created as specified in the testbench, inputs for hardware test may have to use the signals already occurring in the hardware. The idea is for the requirements to describe waveforms or signals that will exist. When this style of requirements is used, hardware testing can take advantage of testing with the signals already present in the circuit. Signals will not have to be injected in

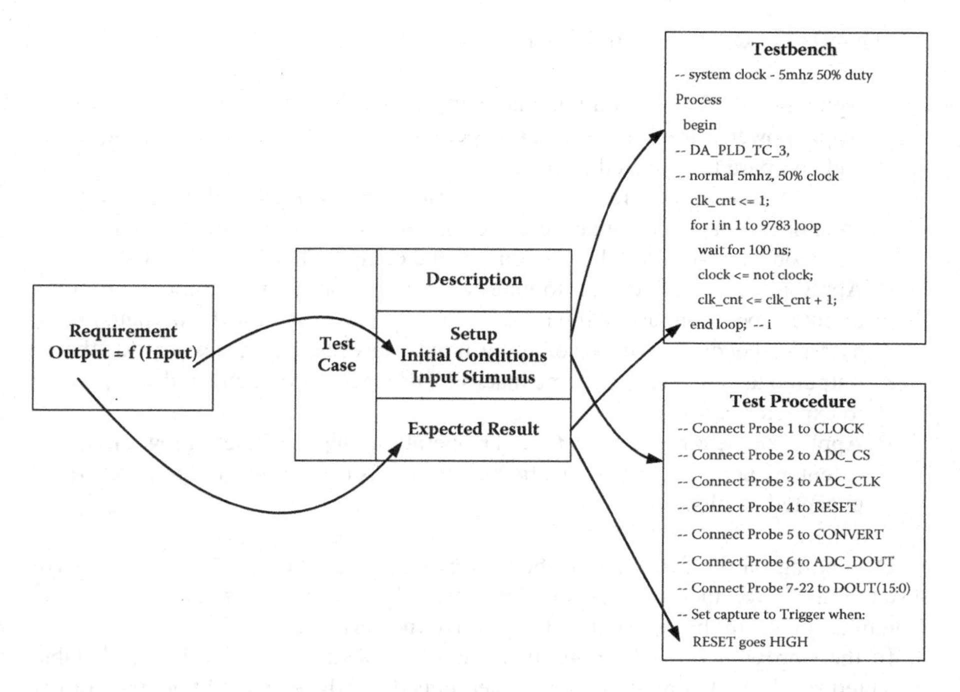

FIGURE 7.15 Requirements and Test Cases

order to achieve requirements-based tests. This style of testing works well for normal range or nominal conditions.

Testing for robustness—with invalid or unexpected inputs—will need signals created and applied to the circuit under test. Hardware tests with test equipment applying inputs to card edge connectors can be configured or programmed to generate invalid inputs. If the signals of interest are on the interior of a circuit, the invalid signals may have to be applied directly to the circuit under test. For PLDs, a stand-alone chip tested allows full access to all input pins and application of any combination of valid and invalid signals to the input pins.

Test cases should be selected for comprehensive verification of the requirement or requirements associated with the test case. The typical sequence for testing is to apply initial conditions and signals, apply power, apply and release the reset, apply the inputs of interest for the test case, and collect the outputs of interest at each step along the way.

Consider a requirements template constructed with the following aspects:

- Power on behavior
- Reset behavior
- Assert conditions
- Deassert conditions
- Response to invalid inputs

The test case will be constructed as follows:

- Apply power, measure output, and compare to expected results.
- Apply power, allow circuit to enter operation, apply a reset, measure output, and compare to expected results.
- Apply power, allow circuit to enter operation, apply a reset, allow circuit to enter operation, apply inputs described as necessary for the output(s) to assert or become active/true. Measure output and compare to expected results.
- Apply power, allow circuit to enter operation, apply a reset, allow circuit to enter operation, apply inputs described as necessary for the output(s) to assert or become active/true, apply inputs described as necessary for the output(s) to deassert or become inactive/false. Measure output and compare to expected results.
- Apply power, allow circuit to enter operation, apply a reset, allow circuit to enter operation, apply invalid inputs. Measure output and compare to expected results.

Depending on the functionality being verified, it may be possible for the fourth test to satisfy (or incorporate) the three preceding tests in the above sequence. Combining tests in this way can help optimize the verification process.

In the step with invalid inputs, there may not always be an easily predictable expected result. If the input clock is varied outside of the expected tolerance of the frequency or duty cycle, there is no guarantee that the circuit will continue to behave in a deterministic manner. In this case, robustness tests can be used to explore and characterize the hardware for how far out of tolerance the inputs can become before the expected behavior no longer occurs. In instances where invalid inputs can be described and applied but the expected result is not known, the test can still be performed. In these instances, the behavior of the outputs should be collected and analyzed by the systems engineer and/or safety engineer to determine whether the behavior is acceptable or if the requirements and/or design need to be updated.

Test cases will test a functional element consisting of a single requirement or a logically related group of requirements. The test case strategy may elect to group coverage for requirements describing a particular output. The test cases will be solely determined from an inspection of the hardware requirements. Design-based tests will not be permitted since they only confirm that the design is the design, and not whether the design meets its intended functionality. Note that if the requirements specify the design's implementation (implementation requirements) rather than its intended functionality, the resulting requirements-based test cases will have the same outcome (and deficiencies) as design-based tests. As noted in the chapter on requirements, this is one of the critical weaknesses of implementation requirements and why they should be avoided.

The effect of design information in requirements on the integrity of the verification process is illustrated in **Figure 7.16** and **Figure 7.17**. As seen in **Figure 7.16**, requirements that express functionality, particularly those written according to the guidance in this book, are equally appropriate for both design and verification, and allow both processes to proceed independently. Since independence is maintained

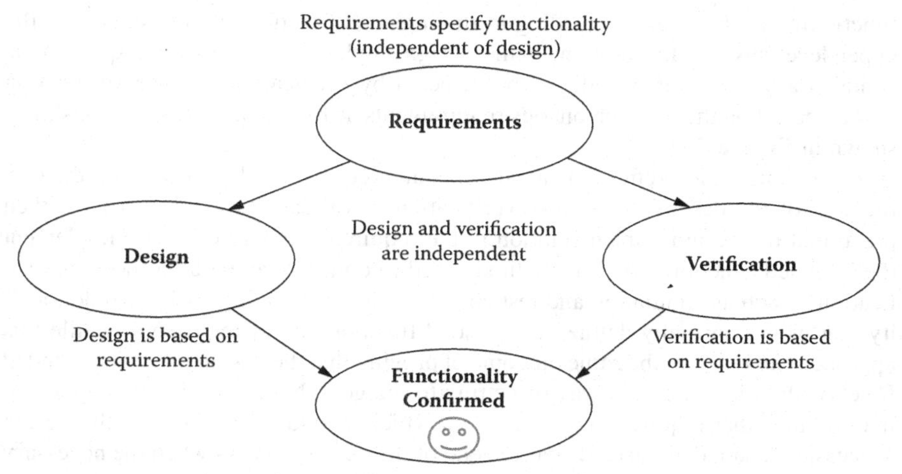

FIGURE 7.16 Functional Requirements and Effective Verification

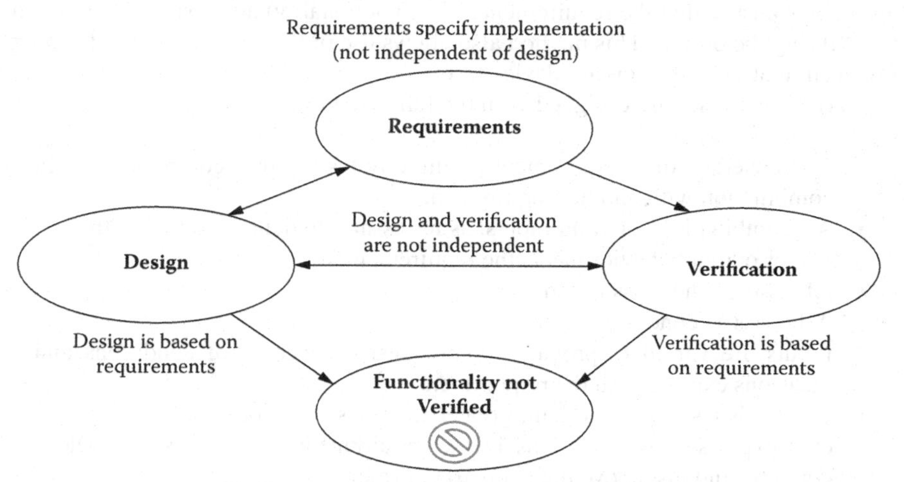

FIGURE 7.17 Implementation Requirements and Ineffective Verification

between the two processes, when the hardware is verified through requirements-based verification, the design is independently proven to comply with the requirements and by extension with the intended functionality as defined by the upper level system.

In contrast, when design information finds its way into the requirements, there is no longer a connection to the system level functionality, no capture of intended functionality, and therefore no way for verification to determine that the design is functioning properly and according to the intended functionality as defined by the upper level system. In addition, if the design implementation in the requirements contains any errors, they will not be detected by verification because verification will be based on the same erroneous requirements as the design. This relationship is shown in **Figure 7.17**.

If implementation requirements are encountered, the burden of proving the correctness of the design shifts from verification to validation. Validation must then prove that the design implementation in the requirements will correctly implement the intended functionality. The methods used to do this may parallel those of verification—such as simulation and test conducted against the higher level functionality that would normally define the intended functionality of the design. While this approach can result in the same outcome of proving that the design meets its intended functionality, it is the hard way of getting to the goal. Note also that this approach may require that requirements validation, which should occur early in the design process, be conducted instead near the end of the design process when the necessary design and hardware are available to validate the requirements. Overall it is not the recommended approach to design assurance.

The test cases are limited to stimulating device or signal inputs and predicting the expected pin level response. Restricting test input and output to the pin or signal level helps ensure that the requirements (i.e., functionality) are verified, as opposed to verifying the design. This method also allows use of the same test cases for functional simulation and in-circuit device tests.

Normal test cases are designed with the following criteria:

- Test coverage of a requirement by directly testing the requirement, or in combination with another requirement
- All combinations of valid inputs, as necessary, to demonstrate that the circuit or output behavior meets the requirement(s)
- All assert conditions tested
- All deassert conditions tested
- Inputs are varied to provide comprehensive coverage of conditions and decisions expressed in the requirement
- Comparisons (e.g., less than, greater than, less than or equal to, equal to, etc.) expressed as conditions in the requirements use inputs just below, equal to, and just above the comparison value
- The effect of the inputs should be observable as a change on the output
- Sequence of tests is considered to show the effect of the inputs on the outputs
- All combinations of valid inputs, as necessary, to demonstrate that the circuit or output behavior meets the requirement(s) and does not have unintended side effects
- Typical input signal timing tolerance
- Typical clock timing tolerance
- Equivalence class of values for input address bus
- Minimum and maximum values for input address bus

- Equivalence class of values for input data bus
- Minimum and maximum values for input data bus
- Input changing before, during, and after clock transition to show registration of input on rising or falling edge as specified in the requirements or design standards
- All possible state transitions are covered when state machines are employed in the design—this will by definition be a design-based test
- Use varying data when the test cases are repetitive
- Use a structured sequence for test cases
  - Apply power
  - Apply reset
  - Allow device to start normal operations
  - Apply desired stimulus
  - Alternate steps in a test case between valid and invalid conditions, assert and deassert behaviors, and always end with a return to valid inputs and outputs in normal operating conditions

Robustness test cases are designed with the following considerations:

- Incorrect inputs or combinations of inputs
- Unexpected inputs or combinations of inputs
- Toggling inputs that are not listed in the associated requirement(s)
- Invalid input timing (e.g., setup and hold violations)
- Invalid state transitions
- Variation of clock duty cycle and/or frequency outside the specified tolerance
- Asserting and deasserting input signals between clock edges
- Applying reset during a test
- Device select/deselect during a test
- Asynchronous or glitch timing
- Inserting additional clock cycles
- Removing clock cycles

Trace tags, such as text with a unique identifier, should be embedded in the test cases so traceability to the requirement associated with the test case can be listed in a trace matrix. Embedding the requirement identifier in the test case can cause an update to the test case when the requirements change, regardless of whether the test cases need to change. The test case author should make every attempt to provide full test coverage of a requirement with a set of test cases in a test case file. Only when absolutely necessary should test coverage of a requirement be spread across multiple test case files (test case groups). The rationale for this grouping is to keep the data together and make the review easier.

Test cases can also use naming conventions for the file names to make it easy to find the data in the configuration management system. An example of the naming conventions follows.

For traceability and ease of comprehension, the following file naming convention can be used.

- Requirement — HRD-XXX-YYY-NNN
- Test case file — TC-HRD-XXX-YYY-NNN.xls
- Test case or step in test case file — TC-HRD-XXX-YYY-NNN _001
  TC-HRD-XXX-YYY-NNN _002
  TC-HRD-XXX-YYY-NNN _003
- Testbench (simulation) — TB-HRD-XXX-YYY-NNN.vhd
- Test procedure (hardware test) — TP-HRD-XXX-YYY-NNN.txt
- Test results log — TR-HRD-XXX-YYY-NNN.log or
  TR-HRD-XXX-YYY-NNN.wav
  (waveform capture)

An example format for a test case is shown in **Table 7.3**.

TABLE 7.3
Test Case Example

STEP	ACTION/INPUT	EXPECTED RESULT
Initial Setup: Steps 1 and 2 initialize the device by setting all inputs to their default states, then toggling the RESET input to clear and initialize the logic.		
1	Set the inputs to their following default states: CLK_50M = 50MHz square wave, 50% duty cycle. AVAL_L = 1. RW = 1. P_ADR(15:0) = 0x0000 RESET = 0	P_DAT(31:0) = 0xZZZZ_ZZZZ
2	Set RESET high for 100usec and then back low to clear and initialize all internal logic to a known state, then wait for 100usec.	P_DAT(31:0) = 0xZZZZ_ZZZZ
Load SDLR_DAT with a known value and then verify that it is cleared to 0x0000_0000 within 1usec after a RESET (CDBR-60(ii)).		
3	Write 0xFFFF_FFFF to address 0x8001 to pre-load the SDLR_DAT register.	None
4	Read address 0x8001 to confirm that the SDLR_DAT register is initialized for this test.	P_DAT(31:0) = 0xFFFF_FFFF
5	Set RESET high for 100usec and then back low. Read address 0x8001 1usec after the falling edge of RESET to confirm that the SDLR_DAT register cleared to 0x0000_0000.	P_DAT(31:0) = 0x0000_0000
...	...	...

Summary of Test Case: This test case confirms that the SDLR _DAT(31:0) data value clears to 0x0000 _0000 within 1usec after the RESET input deasserts low or an SDLR clear command is generated. It reads address 0x8001 both before and after each reset method to confirm that P _DAT(31:0) outputs 0xFFFF _FFFF before the reset, and then 0x0000 _0000 afterward.

Type of Test Case: Normal
Test Case Trace Tag: TC _CDBR-50 _and _55 _and _60 _001
Verification Methods: Simulation, Test
Steps for Test Case:

TEST CASES AND REQUIREMENTS

Now to tie it all together. The requirements express the behavior of the output when the inputs satisfy certain conditions. Using the constructs described in the requirements chapter, the test cases can be readily constructed from the form of the requirement.

EXAMPLE 7.1

An output that asserts when a set of input conditions must be met all at the same time, or the logical AND of the input conditions is expressed:

out1 shall assert high within 50 nanoseconds when the following conditions are satisfied:

- Condition1
- Condition2
- Condition3

The test cases to verify these conditions are shown in **Table 7.4**. The test cases in **Table 7.4** are comprehensive. An exhaustive set of all eight possible combinations of the inputs could also be used. Condition1 through Condition3 could each be a logic one on an input or a more complex expression such as Input1 has remained high for the last 100 nanoseconds, or count100 is greater than or equal to 0X0064.

TABLE 7.4
Test Cases for AND

Condition1	Condition2	Condition3	out1
True	True	True	True
True	True	False	False
True	False	True	False
False	True	True	False

TABLE 7.5
Test Cases for OR

Condition1	Condition2	Condition3	out1
False	False	False	False
False	False	True	True
True	True	False	True
True	False	False	True

EXAMPLE 7.2

An output that asserts when any one of a set of input conditions is met, or the logical OR of the input conditions is expressed:

```
out1 shall assert high within 50 nanoseconds when one or more
of the following conditions are satisfied:

  • Condition1
  • Condition2
  • Condition3
```

The test cases to verify these conditions are shown in **Table 7.5**. The test cases in **Table 7.5** are comprehensive. An exhaustive set of all eight possible combinations of the inputs could also be used. As before, the conditions could be a logic input or a more complex expression.

EXAMPLE 7.3

An output that asserts when none of the input conditions are met, or the logical NOR of the input conditions is expressed:

```
out1 shall assert high within 50 nanoseconds when none of the
following conditions are satisfied:

  • Condition1
  • Condition2
  • Condition3
```

The test cases to verify these conditions are shown in **Table 7.6**. The test cases in **Table 7.6** are comprehensive. An exhaustive set of all eight possible combinations of the inputs could also be used. As before, the conditions could be a logic input or a more complex expression.

TABLE 7.6
Test Cases for NOR

Condition1	Condition2	Condition3	out1
False	False	False	True
False	False	True	False
True	True	False	False
True	False	False	False

EXAMPLE 7.4

An output that asserts when at least one of a set of input conditions are not met, or the logical NAND of the input conditions is expressed:

```
out1 shall assert high within 50 nanoseconds when at least
one of the following conditions are not satisfied:
```

- `Condition1`
- `Condition2`
- `Condition3`

The test cases to verify these conditions are shown in **Table 7.7**. The test cases in **Table 7.7** are comprehensive. An exhaustive set of all eight possible combinations of the inputs could also be used. As before, the conditions could be a logic input or a more complex expression.

TABLE 7.7
Test Cases for NAND

Condition1	Condition2	Condition3	out1
True	True	True	False
True	True	False	True
True	False	True	True
False	True	True	True

TABLE 7.8
Test Cases for XNOR

Condition1	Condition2	Condition3	out1
False	False	False	True
False	False	True	False
False	True	False	False
False	True	True	False
True	False	False	False
True	False	True	False
True	True	False	False
True	True	True	True

EXAMPLE 7.5

An output that asserts when a set of input conditions must be all either be met or all are not met at the same time, or the logical XNOR of the input conditions is expressed:

```
out1 shall assert high within 50 nanoseconds when either all
or none of the following conditions are satisfied:
```

- `Condition1`
- `Condition2`
- `Condition3`

The test cases to verify these conditions are shown in **Table 7.8**. This example uses all eight possible combinations of the inputs to disambiguate the requirements from other logic constructs. As before, the conditions could be a logic input or a more complex expression.

EXAMPLE 7.6

An output that asserts when only one of the input conditions are met, or the logical XOR of the input conditions is expressed:

```
out1 shall assert high within 50 nanoseconds when only one of
the following conditions are satisfied:
```

- `Condition1`
- `Condition2`

TABLE 7.9
Test Cases for XOR

Condition2	Condition3	out1
False	False	False
False	True	True
True	False	True
True	True	False

The test cases to verify these conditions are shown in **Table 7.9**. This example uses all four possible combinations of the inputs to disambiguate the requirements from other logic constructs. As before, the conditions could be a logic input or a more complex expression.

In **Figure 7.18**, the inputs are INPUT1, INPUT2, INPUT3, and RESET. The requirements are constructed as outlined in the requirements section of this book. The power-on behavior is stated in Requirement 1, the reset response is stated in Requirement 2, the assert response is stated in Requirement 3, the deassert response is stated in Requirement 4, and the response to invalid inputs is stated in Requirement 5.

Figure 7.18 shows the requirements for this functional element and lists the test cases associated with each requirement. The test cases are constructed from the requirements. For Requirement 3, INPUT1-INPUT2-INPUT3 all being 1 is verified. Using this test case provides coverage of the "when the following conditions are all satisfied" or logical AND of the inputs. For Requirement 4, INPUT1-INPUT2-INPUT3 all being 0 is verified. Using this test case provides coverage of the "when the following conditions are all satisfied." The test cases for Requirement 5 verify the other combinations of INPUT1-INPUT2-INPUT3. In this case using all eight combinations was a trivial task and would be easy to put into a testbench or test procedure. By organizing the test cases in a table, with the inputs in the columns and the values for the inputs across the rows, verification coverage of the requirements can be readily assessed. **Table 7.10** shows the test cases for the requirements. If one imagines the HDL resulting from this set of requirements, it is also easy to predict that the test coverage of the design from the test cases will yield high coverage. In other words, the elemental analysis will demonstrate that the requirements-based tests provide full coverage of the design.

Note that the test cases can be rearranged and some of the cases reused in order to ensure the visibility of the verification. This sequence is shown in **Table 7.11**. Test cases already defined have been repeated and inserted between other test cases so that the output alternates between high impedance (Z), 1 and 0.

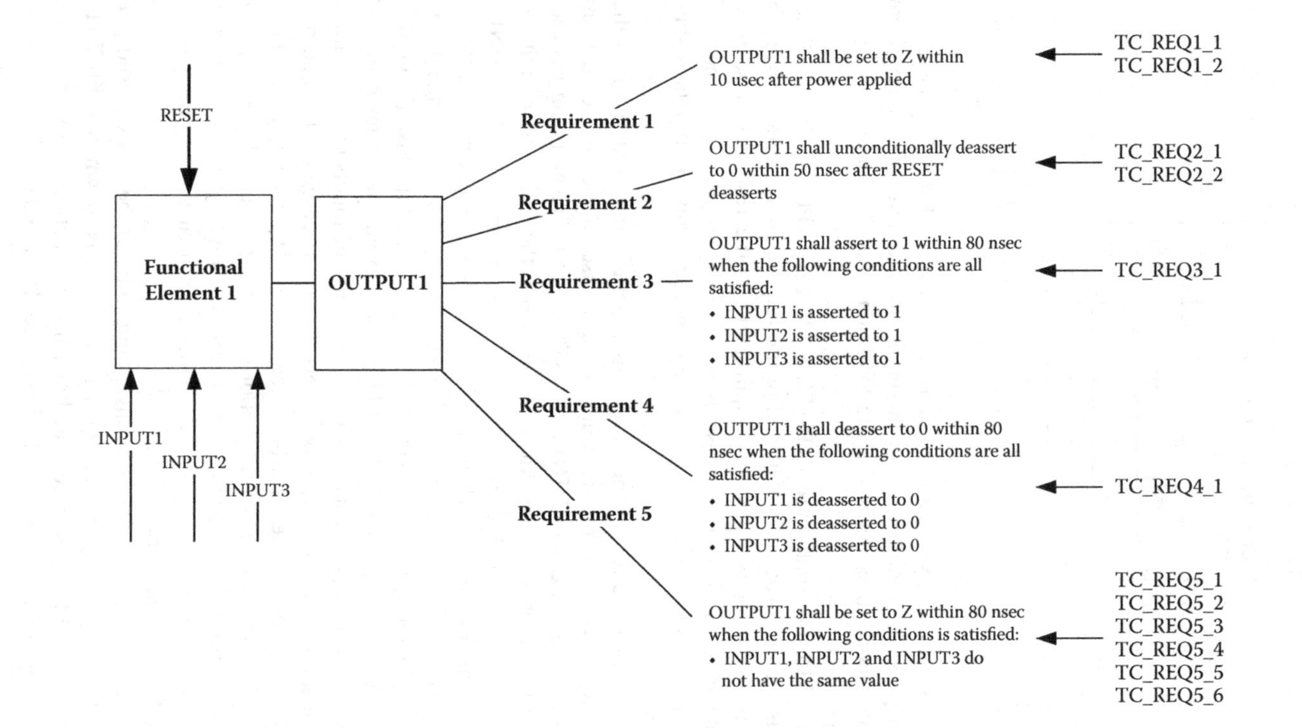

FIGURE 7.18 Requirements and Associated Test Cases

TABLE 7.10
Test Cases for Example Requirements

Test Case	RESET	INPUT1	INPUT2	INPUT3	OUTPUT1
TC_REQ1_1	0	1	1	1	Z within 10 usec after power applied
TC_REQ1_2	0	0	0	0	Z within 10 usec after power applied
TC_REQ2_1	1, then 0	1	1	1	0 within 50 nsec after falling edge of RESET
TC_REQ2_2	1, then 0	0	0	0	0 within 50 nsec after falling edge of RESET
TC_REQ3_1	0	1	1	1	1 within 80 nsec when inputs are all 1
TC_REQ4_1	0	0	0	0	0 within 80 nsec when inputs are all 0
TC_REQ5_1	0	0	0	1	Z within 80 nsec when inputs are not equal
TC_REQ5_2	0	0	1	0	Z within 80 nsec when inputs are not equal
TC_REQ5_3	0	1	0	0	Z within 80 nsec when inputs are not equal
TC_REQ5_4	0	1	1	0	Z within 80 nsec when inputs are not equal
TC_REQ5_5	0	1	0	1	Z within 80 nsec when inputs are not equal
TC_REQ5_6	0	0	1	1	Z within 80 nsec when inputs are not equal

TABLE 7.11
Improved Test Cases for Example Requirements

Test Case	RESET	INPUT1	INPUT2	INPUT3	OUTPUT1
TC_REQ1_1	0	1	1	1	Z within 10 usec
TC_REQ1_2	0	0	0	0	Z within 10 usec
TC_REQ2_1	1, then 0	1	1	1	0 within 50 nsec after falling edge of RESET
TC_REQ2_2_1	1, then 0 then wait 1 usec	1	1	1	0 within 50 nsec after falling edge of RESET then 1 within 80 nsec when inputs are all 1
TC_REQ2_2_2	1, then 0	0	0	0	0 within 50 nsec after falling edge of RESET
TC_REQ3_1	0	1	1	1	1 within 80 nsec when inputs are all 1
TC_REQ4_1	0	0	0	0	0 within 80 nsec when inputs are all 0
TC_REQ5_1	0	0	0	1	Z within 80 nsec when inputs are not equal
TC_REQ3_1	0	1	1	1	1 within 80 nsec when inputs are all 1
TC_REQ5_2	0	0	1	0	Z within 80 nsec when inputs are not equal
TC_REQ4_1	0	0	0	0	0 within 80 nsec when inputs are all 0
TC_REQ5_3	0	1	0	0	Z within 80 nsec when inputs are not equal
TC_REQ3_1	0	1	1	1	1 within 80 nsec when inputs are all 1
TC_REQ5_4	0	1	1	0	Z within 80 nsec when inputs are not equal
TC_REQ4_1	0	0	0	0	0 within 80 nsec when inputs are all 0
TC_REQ5_5	0	1	0	1	Z within 80 nsec when inputs are not equal
TC_REQ3_1	0	1	1	1	1 within 80 nsec when inputs are all 1
TC_REQ5_6	0	0	1	1	Z within 80 nsec when inputs are not equal
TC_REQ4_1	0	0	0	0	0 within 80 nsec when inputs are all 0

8 Process Assurance

Compliance to project plans and standards is ensured by process assurance. Process assurance performs inspections of the hardware and its artifacts to make sure it conforms to its drawings and specifications. Artifacts from the life cycle process are audited to make sure that the processes were followed, and for Level A and B design assurance, that the transition criteria specified in the hardware management plans were met.

While process assurance needs to be performed independently from the audited activities, it does not have to be an independent organization. Companies typically use a representative from the quality organization to perform process assurance activities, which satisfies the independence criteria. A small company can use people in varying roles and meet the independence criteria without a separate quality or process assurance organization.

The activities for process assurance can be performed by personnel that have engineering skills, or personnel experienced with quality assurance responsibilities. The audits and reviews do not require technical design skills, though technical competency can enhance the depth and value of process assurance. The process assurance engineer does not need to perform peer reviews of technical documents. The typical role for the process assurance engineer is to ensure that the peer reviews were performed, records of the reviews were retained in the configuration management system, and that all issues raised in the peer review were resolved.

Process assurance activities should be performed in a timely manner and not left until the end of a project. Using process assurance as a rubber stamp to approve data and hardware at the last minute can compromise product quality, compliance with the project certification basis, and the ultimate certification of the product.

While process assurance persons may not always be design engineers, they are nevertheless valuable team members. Early detection of process issues in a compliance project allows for corrective action to be defined, implemented, and measured. Taking care of deviations in a timely manner will also help ensure ultimate approval of the project.

SAMPLING

Process assurance activities are intended to be performed by sampling the work product. While 100 percent sampling is not required, a smaller sample size should be earned rather than assumed. What this means in practice is that initial audits of engineering work products should look at work from each person on the project.

Taking samples from the most rigorous and conscientious engineer and assuming or asserting that it is a representative sample is not a thorough approach.

For example, when auditing peer reviews, the audits should take samples from each person performing a review and from each type of peer review. Stellar work on requirements peer reviews does not mean that the peer reviews of testbenches were equally proficient.

When trends, positive or negative, are detected in the samples, the trends should be used to justify decreasing or increasing the sample size. Initial audits should be broad, covering many samples of life cycle data and personnel. Once the organization has demonstrated compliant practices, process assurance can reduce the sampling to a smaller percentage. If problems arise later on in the project, the sample size should increase again to determine how widespread the issue is. Once corrective action is implemented and the quality and proficiency in the work product resumes, the audit size can again be reduced.

Audits of work products such as peer reviews, problem reports, baselines, releases, et cetera should encompass any sub-tier suppliers or work outsourced to other organizations.

Note that the sampling concept is not the same for a first article inspection. The first article inspection is a comprehensive review of the production data. Since it is the first production article, it is a sample of the manufactured articles to follow. The manufacturing or production quality control system takes over for the on-going aspects of production to ensure that the subsequent hardware produced is the same as the first article.

CONFORMITY

DO-254 Section 8.1 and especially Section 10.1.6 could imply that process assurance is responsible for conformance or conformity. Process assurance can perform inspections of hardware to ensure that it was built according to drawings and specifications. Conformity for FAA purposes, however, is a formal process performed by an FAA representative or an appropriately delegated individual (or organization). FAA conformity requires a request for conformity, conformity delegation, the conformity inspection, and a statement of conformity. The conformity process is conducted at the system or LRU level which may be the same scope or even higher level of system integration than the aspects covered by DO-254. Project system level certification plans typically describe the plan for conformity and delegation.

Unless otherwise also appropriately delegated by the FAA, the assigned process assurance personnel do not carry out FAA conformity. Process assurance personnel do inspect hardware and associated drawings to ensure that the hardware that will be used in the FAA conformity is manufactured consistent with its drawings.

Process assurance can also perform an in-house conformity for a PLD device. The activity should ensure that the device type and part number is correct, the programming data for programmable devices is the correct part number and version, the programming procedures are documented and followed, and that any device programming verification such as a programming file checksum is confirmed.

AUDITS

Process assurance should perform at least two different types of audits: audits tied to events or activities, and periodic audits. Event specific would include audits performed when phase transition criteria have been met, audits for baseline content, and audits of test readiness. Periodic would include audits of configuration management activities and deviations.

Specific checklists containing the audit criteria are a good method to perform repeatable and documented audits. The Hardware Process Assurance Plan can list the criteria or include the checklists directly or by reference. Once checklists are established for a project, the content and criteria should not be modified without approval from the certification authority.

The hardware process assurance (HPA) audits for a project can be summarized as shown in the example in **Table 8.1**.

Audits should record the configuration identification of all life cycle data inspected. An audit for a peer review should list the data item that was reviewed and the associated peer review checklist. The audit should check whether all paperwork was filled out correctly, peer review actions were completed, and agreed updates to reviewed documents were made in accordance with the peer review comments. The auditor can also look for any additional changes made to the data item that was the topic of the peer review to make sure no additional changes, beyond the scope and intent of the peer review actions, were made.

A spreadsheet of all project audits, their related status, and pointers to the completed audit reports is a powerful tool for organizing and tracking process assurance activities and progress.

AUDITS OF CONFIGURATION MANAGEMENT

All aspects of the Hardware Configuration Management Plan should be audited by process assurance. This includes:

- Verifying that life cycle data have unique identifiers
- Verifying that baseline contents are correct and complete
- Change control procedures are followed
- Problem reporting is conducted in accordance with the plan
- Changes made to life cycle data is reviewed
- Released documents are stored in the correct format in their respective repository
- Periodic backups are performed on servers that host project data and applications
- Off-site storage is used for backup data or disaster recovery
- Data can be retrieved from servers and applications
- Data can be retrieved and restored from off-site storage
- Data in archives is maintained as long as the equipment is used in service on aircraft
- Tapes or disks used for backups or archives are rotated or updated as needed
- Data storage systems are tamper proof

TABLE 8.1
Process Assurance Audits

Phase	Entry Criteria	Activities	HPA Tools	Output	CM Storage	Exit Criteria
Planning	• PHAC, HDP, HVP, HCMP, HPAP released • Hardware standards released	Audit: HPA Planning Completion Checklist	Word, Excel, Visio	Completed Planning Completion Checklist	Planning Completion Checklist HC2 controlled	Planning updates completed
Requirements Capture	Hardware Requirements Document released	Audit: HPA Requirements Checklist	Word, Excel, Visio	Completed Requirements Checklist	Requirements Checklist HC2 controlled	Requirements updates completed
Conceptual Design	Conceptual design of all functional elements complete	Audit: HPA Conceptual Design Checklist	Word, Excel, Visio	Completed Conceptual Design Checklist	Conceptual Design Checklist HC2 controlled	Conceptual design completed
Detailed Design	Hardware Design Data HC2 controlled	Audit: HPA Detailed Design Checklist	Word, Excel, Visio	Completed Detailed Design Checklist	Detailed Design Checklist HC2 controlled	Detailed design completed
Implementation	• Hardware Configuration Index released • Hardware Life Cycle Environment Configuration Index released • Programming file released	Audit: HPA Implementation Checklist	Word, Excel, Visio	Completed Implementation Checklist	Implementation Checklist HC2 controlled	Implementation completed

Formal Test	• Hardware Test Procedures released • Programmed device • Test equipment set up	Audit: HPA Formal Test Checklist	Word, Excel, Visio	Completed Formal Test Checklist	Formal Test Checklist HC2 controlled	Formal Test completed
Production Handoff	• Hardware Configuration Index released • Programming file released	Audit: HPA Production Handoff Checklist	Word, Excel, Visio	Completed Production Handoff Checklist	Production Handoff Checklist HC2 controlled	Production started
All	Ongoing periodic audits	Audit PRs, Change Control Board, peer review records, CM records	Word, Excel, Visio	Completed Periodic Audit Checklist	Periodic Audit Checklist HC2 controlled	
All	Deviation detected	Record deviation and corrective action	Word, Excel, Visio	Completed Deviation and Corrective Action Worksheet	Deviation and Corrective Action Worksheet HC2 controlled	

Some of the above configuration management tasks could be performed by information services or information technology organizations within a company. Process assurance can look at logs of backups or manifests of tapes delivered to off-site storage facilities. Annual or bi-annual visits to archives for an audit can also be performed.

Problem reports should be audited for compliance to the Hardware Configuration Management Plan. Process assurance can look at new problem reports during regularly scheduled change control board meetings. More detailed inspection of problem reports should check that all fields in the problem report are filled in correctly and that the work is progressing toward resolution. The problem report should also be checked to confirm that the proper configuration identification of all life cycle data affected by the problem when the problem is initially detected is recorded. Process assurance can then later check that the configuration identification of all life cycle data updated by the resolution of the problem was recorded. The completion of verification activities such as peer reviews or regression tests cited in the resolution of the problem should also be checked.

SOI AUDIT DRY RUN

Another helpful task for process assurance is to assist in in-house dry runs of FAA SOI audits before the formal audit with the certification authority takes place. The respective SOI audit checklist from the FAA Hardware Job Aid can be used as a guide to make sure all the activities for the stage of development are complete, the associated verification activities have been performed, and all life cycle data is managed in the configuration management system. Corrective actions can be formulated and implemented prior to the authorities discovering problems during the course of the audit. The dry run activity also positions process assurance as a knowledgeable and useful asset in the formal audit.

INSPECTIONS

The most common term for the inspection of hardware at the end of the development life cycle as it transitions to the production environment is First Article Inspection (FAI). The FAI is an examination of the components, assemblies, and their drawings. The first in the First Article Inspection refers to the hardware item being built with the production processes before the serial production cycle begins. In other words, this is the first hardware off the production line.

The FAI ensures that the documentation needed for manufacturing and serial production of the hardware is correct. The inspection also ascertains whether the hardware was built in accordance with the documentation, processes, and procedures.

Airframers and customers may have detailed requirements for FAI. Process assurance should familiarize themselves with customer FAI requirements and help their company adapt and prepare for the event.

DEVIATIONS

Any non-compliance with agreed and approved hardware management plans and standards needs to be first detected, then agreed and tracked. Process assurance is the keeper of the deviation process.

Deviations should be recorded and authorized by engineering and management as required. Corrective actions defined in the resolution of the deviation should be implemented and tracked to ensure that the agreed action is undertaken.

Deviations should also have an agreed escalation process to elevate visibility to management when corrective actions are not defined or not followed. Process assurance should have the autonomy to bring issues to managements' attention without fear of reprisal.

Airframers and customers may have detailed requirements for deviations. Process assurance should familiarize themselves with customer requirements for deviations and help their company comply with the expectations.

SUB-TIER SUPPLIER OVERSIGHT

Work packages and sometimes even whole development projects are sometimes shared with other organizations for schedule and/or budget reasons. The organizations can be another division within a company or a different company altogether. The companies may also have geographic differences—anywhere from a different city to another state or different country.

Process assurance needs to be involved with audits of work done for the project, regardless of the location of the engineers. Some projects may have specific certification requirements for performing on-site audits of work outsourced to another company or division within a company.

When process assurance activities are outsourced to the sub-tier supplier, process assurance at the originating organization still needs to be involved to audit the process assurance at the sub-tier supplier. The process assurance activities at the sub-tier could use the Hardware Process Assurance Plan as the originating organization or have their own in-house Hardware Process Assurance Plan. Regardless of which plan is used or which organization performs the audits and inspections, process assurance at the originating organization will have responsibility and accountability for meeting their DO-254 related objectives.

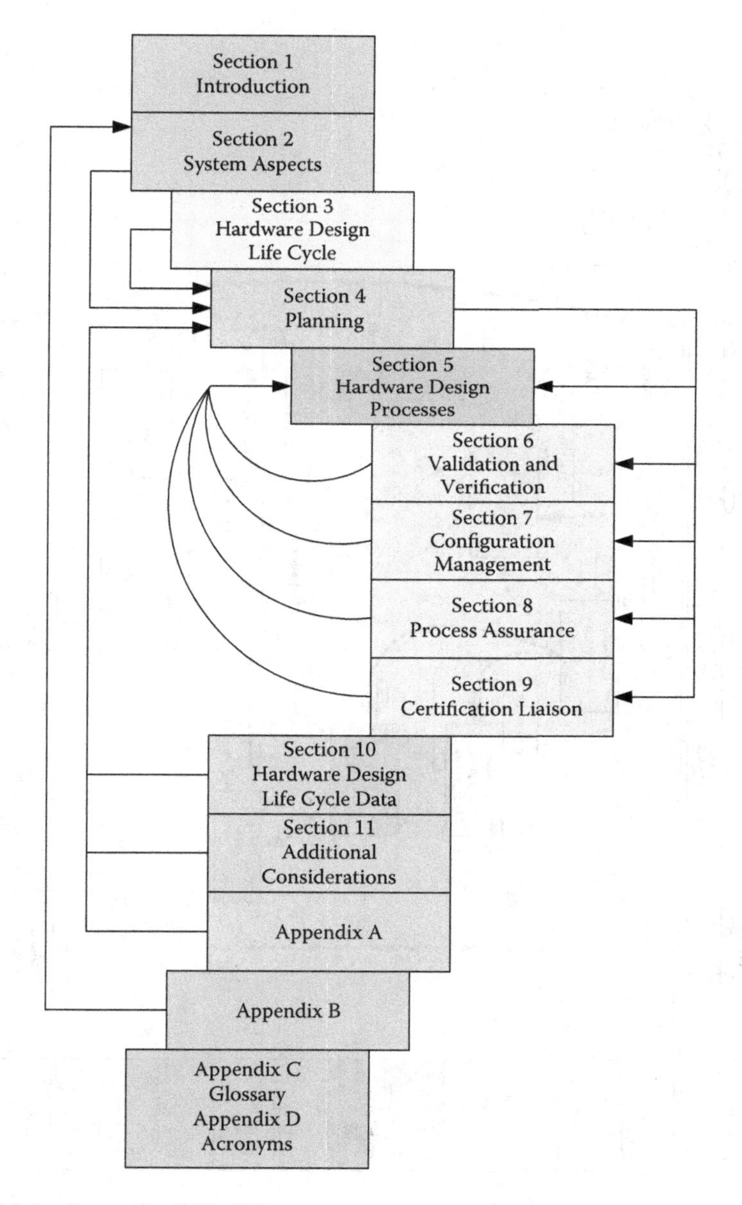

FIGURE 1.1 Structure of DO-254

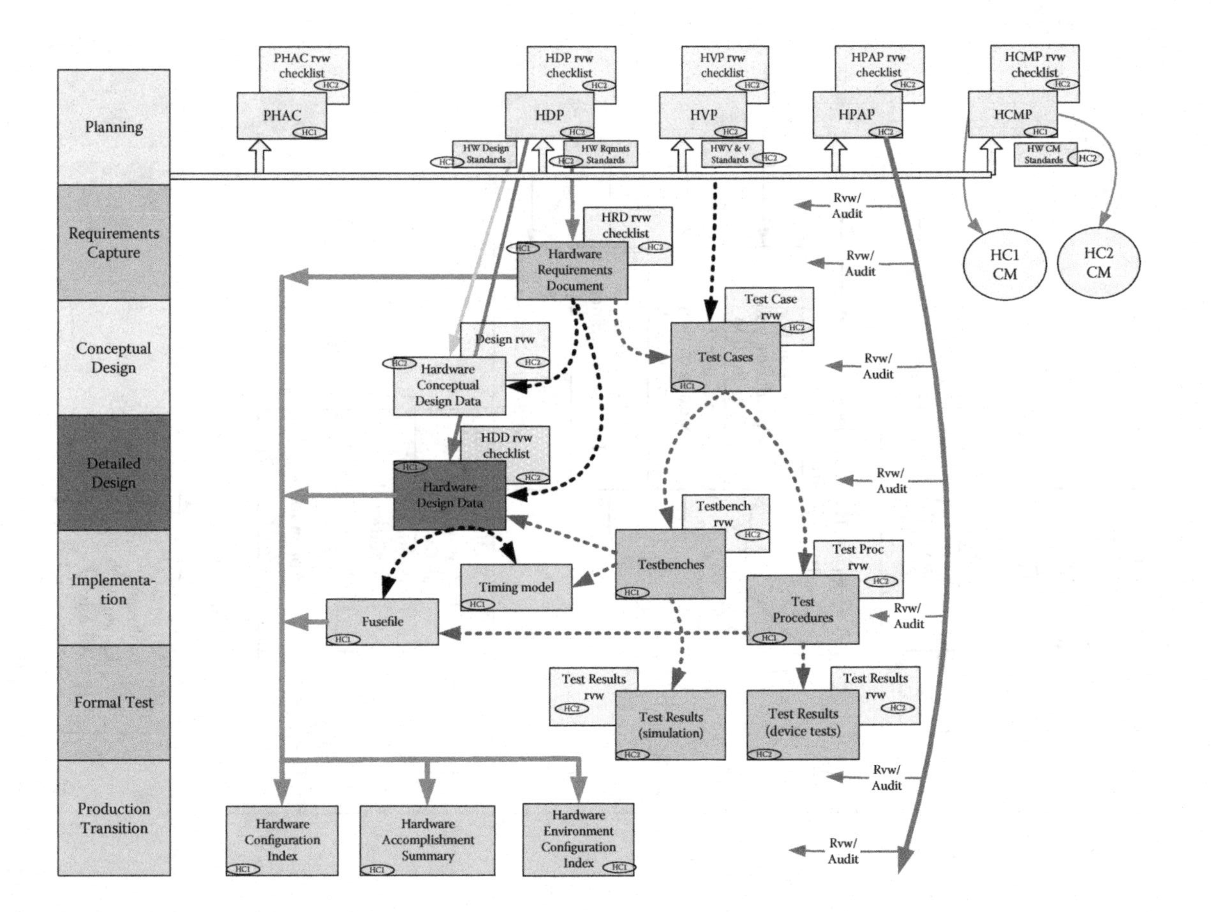

FIGURE 3.4 DO-254 Life Cycle and Data

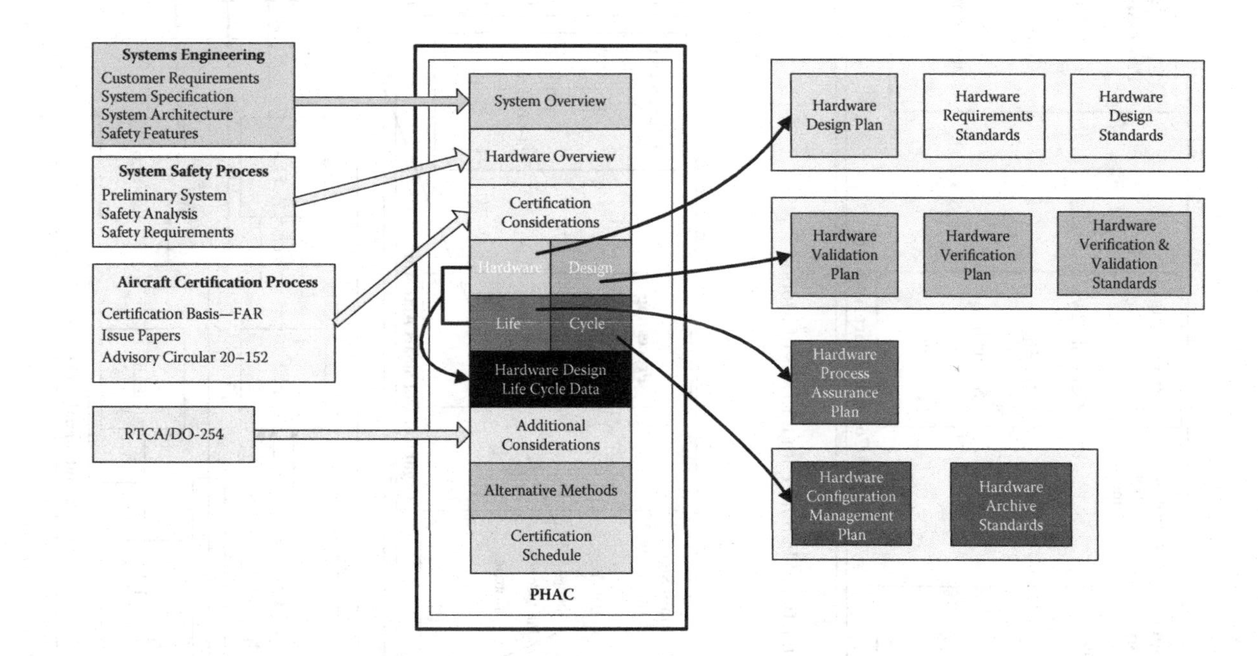

FIGURE 3.5 Role of Plan for Hardware Aspects of Certification

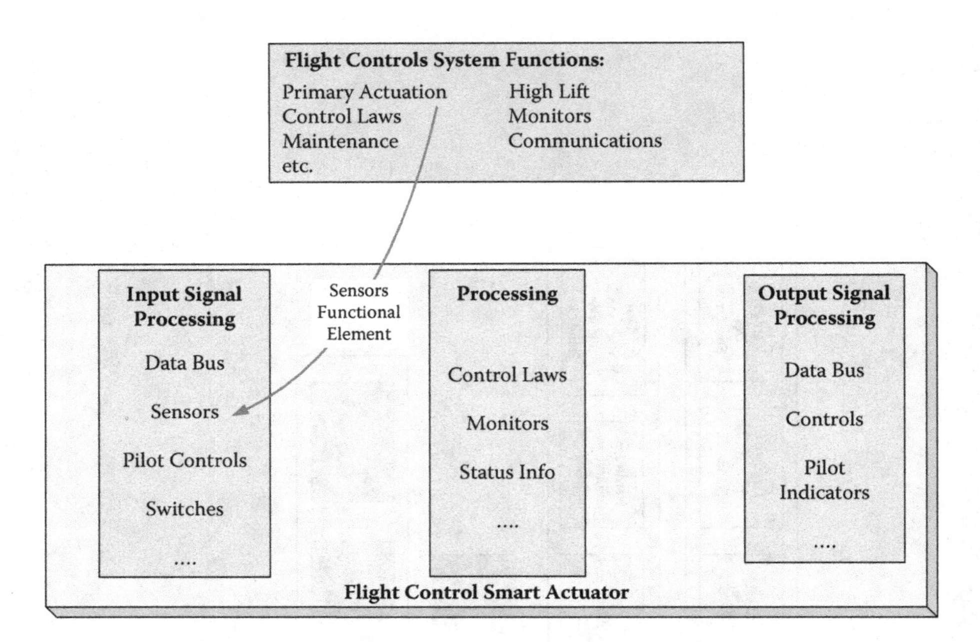

FIGURE 7.2 Functional Failure Path for Loss of Motion

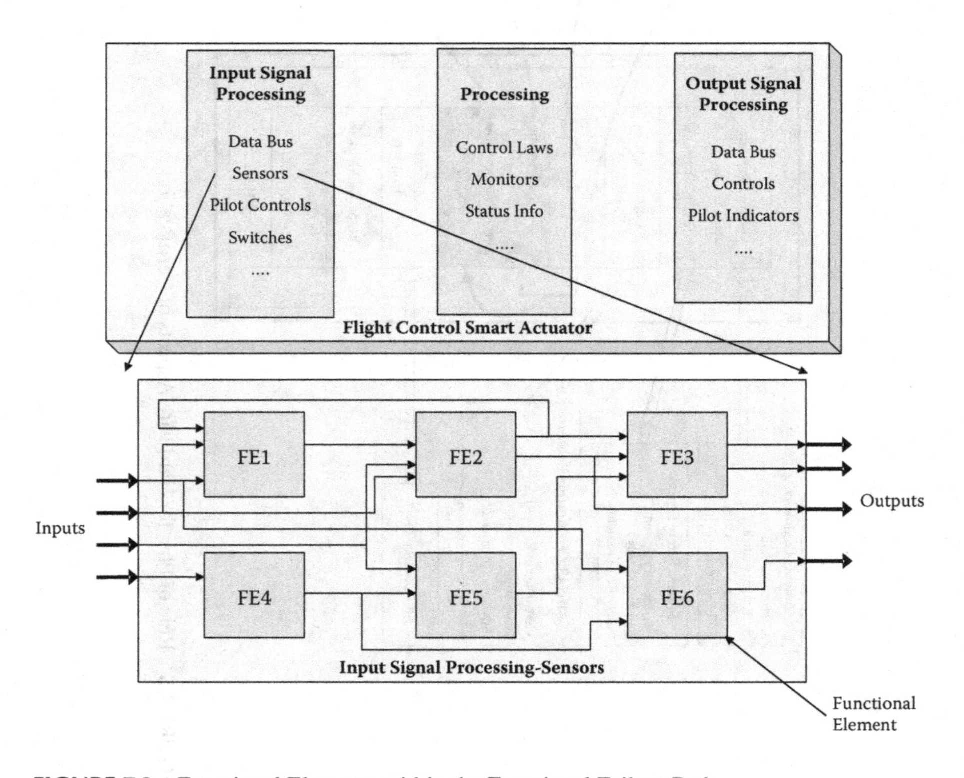

FIGURE 7.3 Functional Elements within the Functional Failure Path

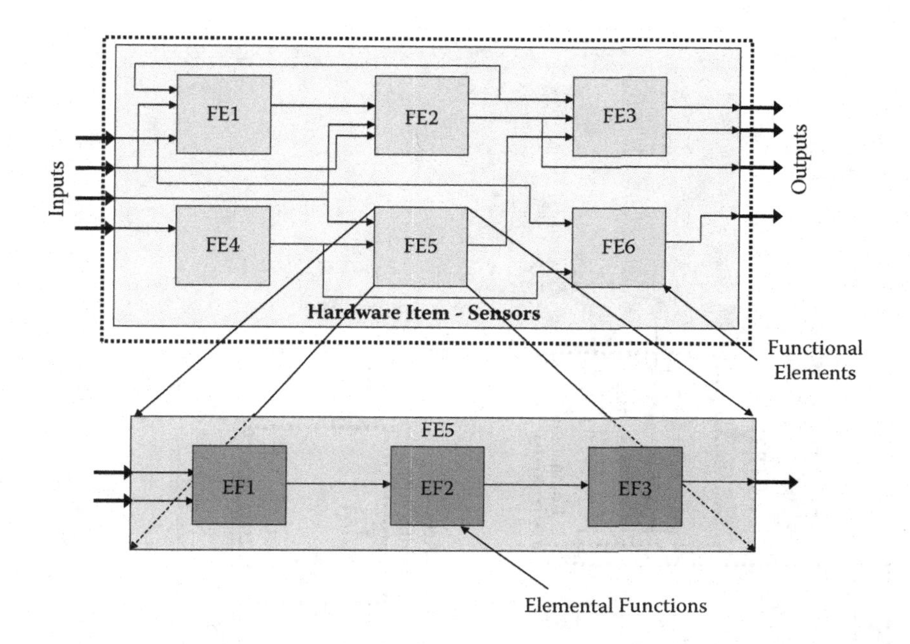

FIGURE 7.4 Elements That Implement a Function

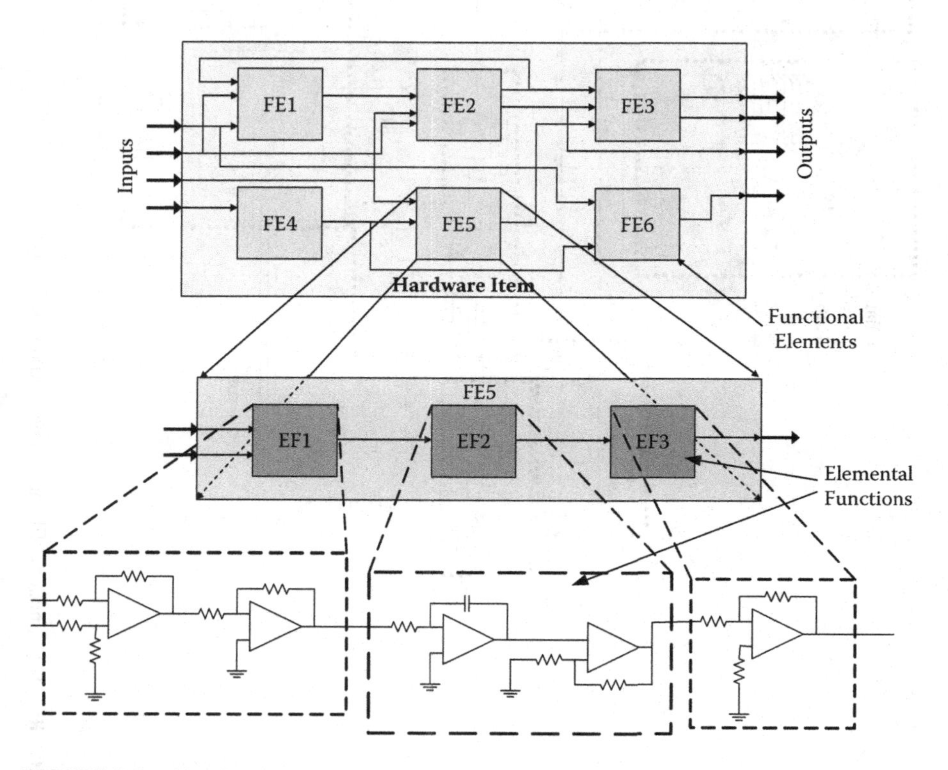

FIGURE 7.5 Circuits within an Element

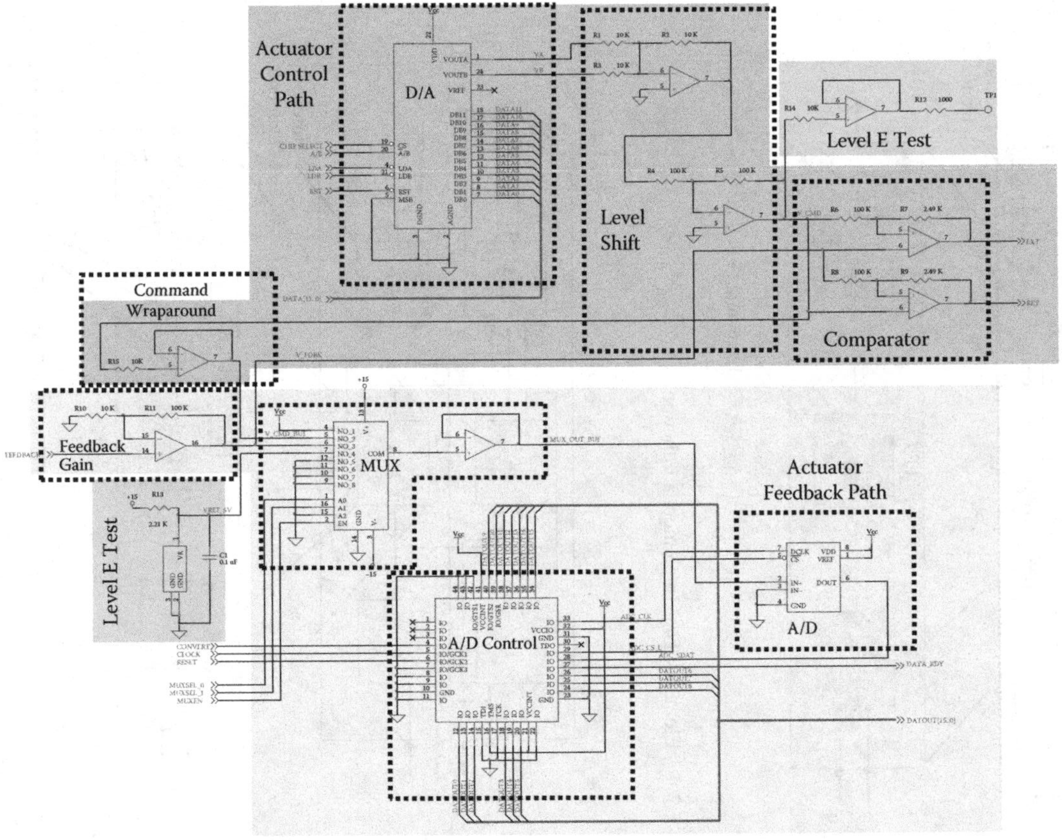

FIGURE 7.8 Functional Elements Overlaid on the Design

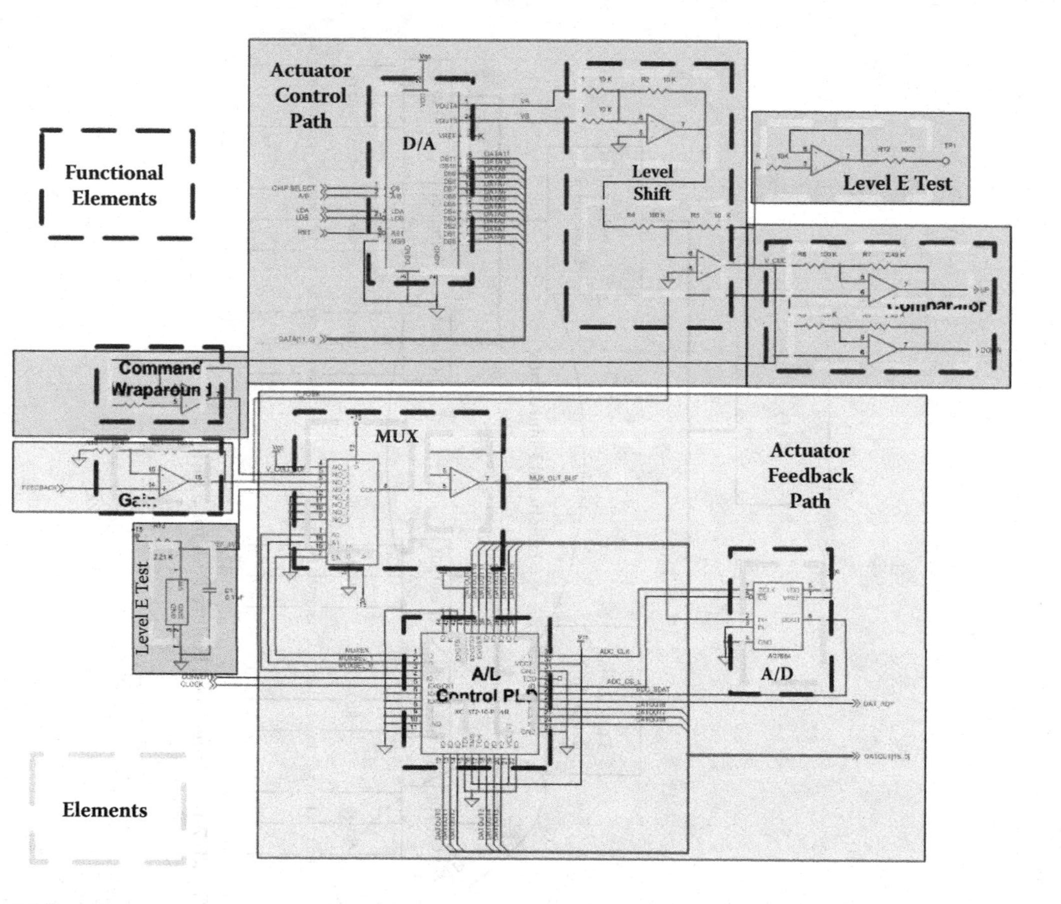

FIGURE 7.9 Elemental Functions

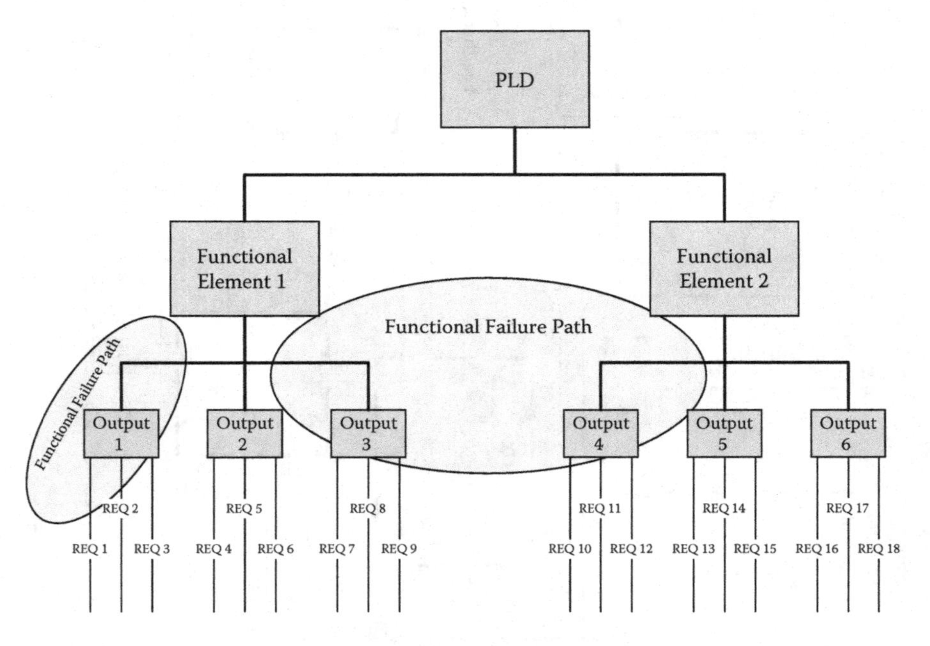

FIGURE 7.10 Functional Failure Path for a PLD

9 Configuration Management

Configuration management is one of the supporting processes of the hardware design life cycle. Configuration management starts during the planning phase, continues through the on-going production of the hardware, and into the long-term archive of life cycle data long after the hardware has ceased production. As with other aspects of design assurance, configuration management starts from the Federal Aviation Regulations: FAR 14 CFR 25.1301 states that installed equipment must be labeled as to its identification. For the purposes of configuration management, the identification of the hardware needs to be unique, that is each system should have a unique part number. FAR 14 CFR 21.31,[1] 21.41,[2] and 21.49[3] lead to the long-term archival of life cycle data, as will be discussed further below.

The configuration management processes allow hardware manufacturers to make consistent copies of hardware items. This could be the manufacture of a line replaceable unit from parts lists and assembly drawings, fabrication of a particular revision of a printed circuit card, applying the programming file to an FPGA, or regenerating the programming file for an FPGA from the source hardware description language (HDL) files and yielding the same programming file checksum.

The configuration management processes also allow hardware developers to make orderly and controlled changes to hardware and hardware life cycle data. This could include modification or deletion of requirements or functions, resolving problems in hardware or hardware life cycle data, or updating versions of hardware installed on a circuit card.

Appendix C of DO-254 has a two-part definition of configuration management. The first part defines configuration management as dealing with configuration identification and the application of changes to items and their identities. The second part defines configuration management as a discipline for identifying and recording functional and physical characteristics of an item, controlling changes made to such characteristics, and reporting the status and implementation of change control. So in this two-part definition, the second part describes how to accomplish the first part.

The definition of configuration management in the introductory paragraph of Section 7 of DO-254 is actually a little more straightforward, and it has the added advantage of defining configuration management through a set of objectives. In plain English, these objectives can be paraphrased as:

(1) Give every configuration item a unique identifier and keep track of them;
(2) Make sure each of those configuration items can be recreated if needed; and
(3) Have a way to control and document every change to every configuration item.

These objectives, regardless of whether they are expressed in plain language or official-speak, are actually fairly simple once the basic concepts of configuration management are understood. These concepts can be summarized as follows:

(1) Each version of each item of life cycle data (files, documents, design data, verification test cases, test procedures, test results, etc., including collections of items) must be uniquely identified such that no other piece of data could ever be confused with it.
(2) Each uniquely identified configuration item must be documented with a paper trail (physical or electronic) that will allow it to be located whenever it is needed.
(3) Each piece of life cycle data must be kept in an environment that is safe from loss.
(4) Each piece of life cycle data must be protected (the level of protection depends on the data type and DAL) from unauthorized changes. This protection can be procedural, electronic, or both.
(5) Changes must be tracked and documented (as with protection, the level of tracking and documentation depends on the data type and DAL), and only the changes that are documented should be made.

WHY CONFIGURATION MANAGEMENT?

Most engineers learn through first-hand experience how hard it can be to keep track of the versions of a design or document, and how painful it can be when versions get mixed up. The pain and difficulty are compounded when the pace of development is fast and multiple versions of an item are being worked at the same time. Ideally, a single bad experience is enough to convert an engineer to the gospel of configuration management.

When engineers are first introduced to formal configuration management, they may not fully appreciate the benefits, which can lead to resistance. This is both common and normal for engineers who are either new to the idea of configuration management or who have never experienced the cost of configuration errors; engineering is well known to be much more fun when unfettered by processes and activities such as configuration management. However, unrestrained development, while fun and usually faster as well, is not appropriate for high integrity design processes because of the higher probability of errors. As noted in the Introduction to DO-254 chapter of this book, it is the supporting processes such as configuration management that allow a design process to achieve and maintain the level of integrity that is requisite for airborne electronic systems.

Configuration management then is the supporting process that minimizes the probability of errors due to incorrect identification of configuration items and their versions, lost or misplaced configuration items, unauthorized changes, incorrectly implemented changes, and inadequately tracked issues. As such, it provides the following for a project and its data:

- Tracking and uniquely identifying each version of a configuration item.
- Provides a known state of the configuration item for development (including verification), delivered hardware, and production.
- Provides for repeatability for every version of the configuration item.
- Provides confidence that the configuration item is the right item, and that the configuration item is built right.
- Ensures that problems or issues with a configuration item are identified, tracked, documented, and properly resolved.
- Ensures that changes made to the configuration item are properly managed and controlled, including:
  - Only authorized changes are made.
  - The correct version of the configuration item is changed.
  - The changes are tracked from beginning to end.
  - The changes are made in a deliberate and controlled fashion.
  - The changes are implemented as intended.
  - The changes meet their intended goals.
- Ensures that every version of the configuration item is safely archived.
- Ensures that every version of the configuration item can be retrieved or recreated.

While not required by DO-254, an electronic file management system with built in version control can help organize project data. These systems often allow groups or sets of data to be tagged and retrieved as a collection. These features support the formation of baselines and the hardware build procedures. File management systems often allow back-end customization and integration with a problem reporting (change request) tool that ties changes to the specific version of a configuration item. Many such electronic tools have reporting features and search capabilities that help project management aspects. A standard project file directory structure can be set up so that new projects simply need to populate data as the project progresses. It is recommended that data be organized by: (1) project; then (2) the version of the release or baseline; and then (3) by configuration item(s) within each release. Given the large volumes of data that are generated during a DO-254 project, it behooves an organization to use tools to store and manage the data.

For traceability and ease of comprehension, a standard file naming convention can be used. The naming convention can be used in conjunction with the electronic file management system. An example of this type of data organization follows:

• Requirements document	`HRD-13579-RevA`
• Requirement	`HRD-123-456-789`
• Test case file	`TC-HRD-123-456-789.xls`
• Test case in test case file	`TC-HRD-123-456-789 _001`
	`TC-HRD-123-456-789 _002`
	`TC-HRD-123-456-789 _003`
• Testbench (simulation)	`TB-HRD-123-456-789.vhd`

- Test procedure (hardware test) `TP-HRD-123-456-789.txt`
- Test results log `TR-HRD-123-456-789.log` or `TR-HRD-123-456-789.wav` (waveform capture)
- Requirements peer review `PeerRvw-HRD-13579-RevA.doc`
- Test case reviews `PeerRvw-TC-HRD-123-456-789.xls`
- Testbench reviews `PeerRvw-TB-HRD-123-456-789.vhd`
- Test procedure reviews `PeerRvw-TP-HRD-123-456-789.txt`
- Test results reviews `PeerRvw-TR-HRD-123-456-789.wav`

DATA CONTROL CATEGORIES

DO-254 recognizes that some configuration items are more critical to design assurance than others by defining two data control categories. These data control categories are essentially the same as those defined for software in DO-178, but adapted to hardware and its processes. In fact, inspection of the configuration management portions of DO-178 will show that there are significant similarities between hardware and software configuration management and activities.

Configuration items that are more critical to design assurance, and which therefore require tighter configuration management and control, are defined as being hardware control category one (HC1). Configuration items that are less critical, or in other words that will have less effect on design assurance if it becomes corrupted in some way, are defined as being hardware control category two (HC2).

CONFIGURATION MANAGEMENT ACTIVITIES

Section 7.2 in DO-254 describes the activities that may be used to achieve the configuration management objectives. As with all of the activities described in DO-254, the configuration management activities are recommendations on how to achieve the objectives, but are not required, nor are they the only acceptable way to achieve the objectives. However, when considering the objectives of configuration management against the basic concepts of what configuration management is and should accomplish, the activities in DO-254 are arguably the most reasonable means of implementing configuration management.

Five configuration management activities are defined in 7.2: configuration identification, baselines, problem reporting/tracking, change control, and release/archive/retrieval. Each of these activities is elaborated in sections 7.2.1 through 7.2.5 of DO-254.

Rather than analyze the activities in sections 7.2.1 through 7.2.5, a more useful approach for this discussion of configuration management is to examine Table 7-1 of DO-254. Table 7-1 identifies the specific configuration management activities (taken from sections 7.2.1 through 7.2.5) that must be applied against HC1 and HC2 configuration items. This table contains a great deal of information on the application of configuration management activities, and every DO-254 practitioner should make a point of carefully studying this table and its referenced paragraphs to understand exactly how configuration items must be managed. An understanding of DO-254

Table 7-1 will go a long way toward mastering the fundamental concepts and practices of configuration management.

DO-254 Table 7-1 lists 11 specific configuration management activities, references the DO-254 paragraphs that define them, and identifies the hardware control categories to which they apply. Or alternatively, since all 11 activities apply to HC1 data, the table defines whether each activity applies to HC2 data as well. Thus from DO-254 Table 7-1 it can be seen that the difference between HC1 and HC2 data is that HC1 requires the additional activities of baselines, problem reporting, records/approvals/traceability of changes, release, and archive media selection/refreshing/duplication.

CONFIGURATION IDENTIFICATION

Configuration Identification is the activity whereby each configuration item is unambiguously identified. This unambiguous identification means to assign each configuration item a unique identification (such as a name, number, or both) that identifies both the item and its version. As noted earlier, this identification (ID) must be unique, meaning that no other configuration item can have the same ID. The dissimilarity in configuration identification can take many forms, but when using most of the commercial computer-based configuration management tools it boils down to giving each data item a unique name (or number) and letting the configuration management tool preserve and manage the different versions of the item. In the configuration management tool the configuration item will normally look like a single item that has all of its versions stored in the tool's database rather than appearing as multiple instances of the item that can all be seen at once. Alternatively, each version of the configuration item can be maintained as separate entities that are each assigned a unique name. Regardless of which approach is used, each version of each configuration item must be uniquely identified so that no two instances of any data items have the same ID.

Provisions should be made to provide unique configuration identifiers for each type of configuration item. While documents, drawings, and schematics traditionally have naming schemes and ensure a unique identity, the large number of other files and data in a DO-254 project also need to have unique names. The following list details the various types of configuration items that require unique identifiers:

- Plans (hardware management plans such as PHAC, HDP, etc.)
- Standards (requirements, design, archive and verification/validation)
- Requirements document
- Design document
- Design data—schematics, layout drawings and files, parts list, assembly drawings
- HDL files
- Programming files
- Test cases
- Test procedures
- Test results (hardware test)

- Testbenches (simulation)
- Test results (simulation)
- Trace data (if separate from requirements, design, or verification documents)
- Peer review forms (templates)
- Completed peer review forms
- Process assurance audit and review records

When using a file management or version control tool, the namespace (i.e., a unique name) is often used for ensuring that a file has a unique identification. In this case, the file name should include the complete directory specification of the file so that identically named files used on different projects can be distinguished. In general, file names can be project specific to help provide a unique identity. But for cases where a common name enhances the comprehension of a design, such as a `top.vhd` file, then the complete file name should be specified as the identifier. Files reused on another project can have the same file name with a new directory path.

It is not recommended that two configuration items be given the same name and then be stored in different locations on a workstation, laptop, or personal computer that is outside of a file or version control tool.

BASELINES

Baselines are like taking a snapshot of a configuration item and making that snapshot a formal version (configuration) of the item that is then given a unique configuration ID. Baselines should be established for each significant version of a configuration item, meaning each version that will be used for some form of certification credit. For example, a document should be baselined (stored in a configuration control system) for each formal peer review so there is no confusion or question about which version of the document or configuration item was reviewed. Any changes that must be made to the document or configuration item as a result of the review are then made to the same version of the document that was reviewed, ensuring that the review and update process maintains its integrity.

Baselines can also be implemented for groups of data, such as a group baseline of all of the requirements and traceability for all hierarchical levels of an LRU (LRU, circuit card, and PLD requirements, plus the requirements traceability between them). Note that in this example, creating a baseline of the HC1 item (the requirements) includes HC2 data as well (the requirements traceability), so while baselines are created for HC1 data, some of those baselines can include HC2 data.

BASELINE TRACEABILITY

Baseline traceability, which applies to both HC1 and HC2 items, is the activity in which two baselines of the same item are linked together, or in other words there is a documentation trail between the two baselines. This traceability should be adequate to establish the lineage or descent of the new baseline from the previous one, including the explicit documentation of any changes that were implemented when going from one to the other. As a goal, the changes captured in the traceability should be documented well enough to enable either baseline to be recreated from the other. DO-254 section 7.2.2-4 states that this traceability is one aspect of claiming certification credit for the design assurance artifacts from a previous baseline.

According to DO-254 Table 7-1, baselines apply only to HC1 data items, but baseline traceability applies to both HC1 and HC2. This apparent contradiction—after all, how can there be baseline traceability for HC2 items when HC2 items do not use baselines—is explained by Footnote 1 in Table 7-1, which states that just because an HC2 item is used as part of a new baseline does not mean that it has to be elevated to an HC1 item. What this means is that new baselines that contain HC2 data (such as the traceability in the group requirements baseline example above) have to include traceability between the old and new versions of both the HC1 and HC2 data items in both baselines. So while HC2 items are not subject to baseline traceability when taken as single items because HC2 items by themselves are not baselined, they do have to be traced when they are baselined as part of an HC1 item.

Baseline traceability can be established on two different levels. The first level is through a change history embedded in or associated with an HC1 configuration item. For instance, a PHAC can have its original release—Revision A. The next release is Revision B. The change history in Revision A of the PHAC will simply state that the document is being released for the first time. The change history in Revision B of the PHAC will list the changes (or reference the problem reports or change requests) to create Revision B from Revision A. The traceability between Revision A and Revision B of the PHAC is established through the change history and the problem reports or change requests.

Baseline traceability for an aggregate group of configuration items is established with a configuration index. The first release of the configuration index could be, for example, the –001 version of a PLD. The revision history of the configuration index will simply state that the document is being released for the first time. The change history section of the configuration index will state that the –001 version is the first release of the hardware. The next release of the configuration index would be for the –002 version of the PLD. The change history in Revision B of the configuration index document will list the changes (or reference the problem reports or change requests) to create Revision B from Revision A. The change history section of the configuration index will list the problem reports or change requests that were incorporated in –002 version of the hardware.

Baseline traceability for an aggregate group of configuration items such as an LRU can be established with the top level drawing in a manner similar to a configuration index for PLDs.

PROBLEM REPORTS

Problem reporting is the formal process through which problems or changes to a HC1 controlled configuration item are documented, tracked, and resolved. A problem report is generally a structured record that is maintained electronically in a problem reporting system, but other ways of tracking problems are acceptable as well as long as the system has the necessary integrity to ensure that problems are tracked from discovery to resolution. However, whether the problem reporting system is implemented with an electronic tool or through index cards, its integrity will ultimately depend on the quality of the problem reporting procedures and whether the people who use them will conscientiously follow those procedures. Problem reports can also include change requests—changes that do not stem from an actual problem.

Problem reports are required for HC1 configuration items once the item is released. When a problem is discovered it should be immediately recorded in a problem report so the problem can be tracked to its eventual resolution. The problem reporting system should track problem reports (and therefore the problem itself) from initiation to closure to ensure that the problem is being worked and that it is eventually resolved.

Problem reports can also be an integral component in the change control activity, since they are an ideal mechanism for approving, documenting, and tracking changes to an item.

A sample work flow with the various phases of completion of a PR is shown in **Figure 9.1**.

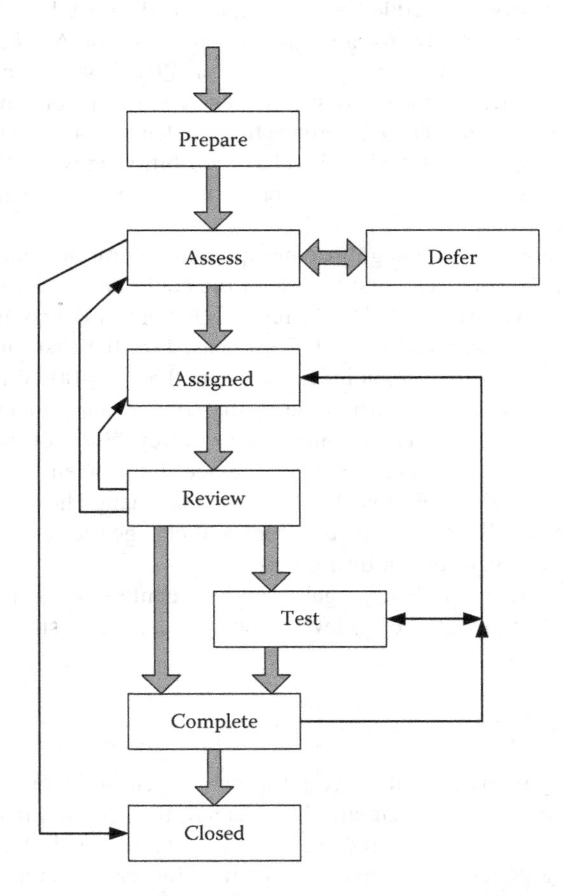

FIGURE 9.1 Problem Report Work Flow

Problem reporting, regardless of the system or level of automation used, should adhere to some basic principles:

- The description of the problem should include all the information needed to accurately reproduce the problem
  - The hardware item, including part number and serial number, if a hardware failure or error is reported
  - The configuration identifier and version of all life cycle data involved in describing the problem
    - Plans
    - Standards
    - Peer reviews
    - Test procedure (if a test was run and failed)
    - Test results that identify the problem
    - Simulation results that identify the problem
- The PR should include a description of how the problem was found
  - Through a review
  - During a test
  - Reported by a customer
- A root cause analysis that identifies the exact nature of the problem or error. The description of the analysis should be detailed enough to allow the analysis to be independently reproduced or reviewed.
- A change impact analysis that identifies the changes required for the hardware and/or life cycle data. The change impact analysis should clearly describe the change needed for each piece of life cycle data and the activities that need to be repeated, such as rework of peer reviews or a rerun of tests or simulations.
- A detailed description of the changes that were actually made, or a reference to where that information can be found.

The HCMP should define the phases or process flow for problem reporting. All data recorded in the problem report for each phase should also be clearly described. The administration of problem reports is usually performed by a change board. The change control board reviews new problem reports, ensures that the problem reporting system is being used correctly, determines when changes will be incorporated, authorizes changes or solutions to problems, and finally closes out problem reports when all is done. The change control board should also ensure that any problem reports deferred beyond the compliance approval are adequately analyzed for system level impact and determination of potential aircraft safety impact.

The PR starts with a complete description of the starting point of the problem. This should include the item identifier and revision of all life cycle data. The description should be complete enough to allow the problem to be reproduced. For example, if the problem is conflicting requirements, the PR would state the requirements document number and version, the identifier of the two requirements, and a description of the discrepancy. The problem description should be limited to describing the problem, and not contain information pertaining to the analysis and root cause, or to the solution or fix for the problem. If the problem is a test failure, then the PR should identify the hardware item under test, the test procedure that was run, and the test results obtained.

The PR is assigned and an analysis is performed to determine the exact nature of the problem and the underlying fundamental root cause. Once the root cause is known, the analysis can then expand to identify all life cycle data and activities that are impacted or invalidated by the problem.

The change impact analysis should describe the change needed to the hardware and the life cycle data. The analysis should also state which verification activities need to be repeated.

The closure phase of a PR should clearly list the final version of all life cycle data that was updated or reworked as a result of resolving the problem, and how the items were updated or reworked since the actual fix may not be precisely as described in the change impact analysis. This may include the requirements, design, test cases/procedures/benches/results, and all associated peer reviews. Some problems could extend as far as a rework of plans and standards as well as all subsequent life cycle data.

Recording and tracking how problems were found is an excellent way to identify how well the verification activities are performing. If peer reviews rarely find problems and all the problems arise in the test phase, then more training and emphasis could be put on the review process to catch and eliminate problems earlier in the life cycle.

DO-254 section 7.2.3 lists only two types of data that should be documented in problem reports: the configuration of the affected item (the complete identification of the item and its version), and a description of the action taken to fix the issue. However, that does not mean that those are the only types of information that should be recorded in a problem report. If using an electronic problem reporting system, merely documenting an issue and its resolution underutilizes much of the potential of such systems. To use a problem reporting system to its greatest potential, the problem reports can be used to document all aspects of the problem from discovery to resolution. This approach to problem reports has a number of benefits and allows the problem reporting system to be used in ways well beyond its basic purpose of reporting and tracking problems.

Consider the possibilities when problem reports contain the following information:

- A headline or title that contains keywords identifying important aspects of the problem, such as the project, the hardware item's name and part number, the type of problem, how the problem was discovered, and a very brief description of the problem. Headlines filled out in this manner will make identifying problem reports significantly easier and faster, as well as make it possible to discern the topic of each problem report without having to open and read it. The keywords can be used with filters or searches to gather statistical data on problems and their characteristics, allowing users to identify trends and problem areas that would otherwise escape detection. They can also be used to locate all problem reports associated with each project, hardware item, problem type, etc., to gather statistics on performance and reliability for individual items. When creating standards for filling out problem reports, consider creating a standardized headline format with standardized keywords for the types of statistical data that may be valuable in the future, including post-project lessons learned.

- A problem description that documents everything that is known about the problem. This information should be detailed and comprehensive enough to allow someone other than the author to conduct an analysis and recreate the problem without having to ask the author for more information. This part of a problem report should include the explicit identification of the item that had the problem (including its name, part number, version, and location in the configuration management system, if applicable), what happened (including the symptoms of the problem), how the problem was discovered (including all relevant details about the conditions under which the problem appeared), and how to reproduce the problem. As a rule of thumb, if anyone has to ask the author questions about the problem, then the problem description is probably not thorough enough. The problem description should not contain a description of the analysis of the problem nor the solution—those should be documented elsewhere in the problem report to keep information partitioned according to topic. If the problem description is too large for the problem report's description field, summarize the problem in the description field of the problem report and put the detailed description in a file that is then attached to the problem report. Most electronic problem reporting systems allow documents and other files to be attached; if this feature is not included in the problem reporting tool, or if a non-electronic system is used, the problem description should be placed in a configuration management system and referenced from the problem report.
- A description of any analyses that were conducted to reproduce the problem and determine its root cause. The analysis should start with the information in the problem description and from there develop enough additional information to allow the problem and its root cause to be understood, and to allow potential solutions to be developed. The description of the analysis should be detailed and thorough enough to allow a reader to understand how the problem was analyzed, how the analysis was conducted, what the analysis revealed (including the root cause, if known), how to recreate the analysis if needed, and potential solutions to the problem. It should also analyze the impact of the issue and its resulting changes on all related items and processes, and identify and document all of the other items, documentation, and processes that will be affected as part of the ripple effect of the initial problem and its resolution. This includes any feedback that may be needed to other processes and activities. However, how the problem was actually fixed should not be documented in the description of the analysis, nor should the analysis only describe the possible solutions.
- A description of the resolution or disposition of the problem. This information should describe how the final solution was determined and how it would solve the problem, a detailed description of how it was implemented, the complete identification of the item and version in which the solution was implemented, and how the efficacy of the solution was independently verified.

The reason for including so much detail in the problem report is twofold: first, skimping on details will only increase the cost of the problem report as well as increase the chance of errors by forcing every reader to search elsewhere for the information they need; and second, putting every known fact into the problem report will make it a self-contained history of the problem, which means all information about the problem and its resolution will be documented in a single location.

CHANGE CONTROL

Change control has two components: integrity and identification; and records, approvals, and traceability.

The first component of change control—integrity and identification—applies to both HC1 and HC2. This activity ensures that the integrity of a configuration item is preserved when changes are needed by ensuring that the item is changed only when actually necessary. This is accomplished through two mechanisms: first, each potential change to an item should be assessed to determine whether the item truly needs that change, and second, the item should be changed only when authorized. The end result of this activity is to make sure that all changes to a configuration item have been assessed to ensure that changing the item is acceptable in all respects.

Assessing changes to determine whether they are actually needed is a way to minimize the number of updates to an item. Updates can be time consuming and expensive for HC1 items, and for both HC1 and HC2 items, frequent or unnecessary updates can cause configuration problems if they are not managed well. It makes sense that all items should be updated only when necessary. The measurement of what is necessary will vary depending on the item, and there are no standardized guidelines on when a change should be implemented as opposed to deferred, but in general if a change will not alter the fundamental characteristics of the item then it is probably best deferred until a later date. For example, a typographical error that will not affect the linguistic or technical content of a document can be left in queue for a future release instead of forcing an immediate update.

Preventing unauthorized changes makes good sense for any type of information, and is essential for controlling the configuration of data. This applies to both HC1 and HC2 items, and while DO-254 provides no guidance on who should authorize changes nor on the criteria that should be used in the decision, the unwritten expectation is that the level of control in this activity should be considerably more stringent for HC1 items than it would for HC2. For HC1 items the authority to make a change is implicit in the approval of the change (part of the second component of change control, which applies only to HC1 items), so for HC1 items this activity can be tied to the problem reporting or change control activity by requiring that a problem report or change request be initiated for any change to the item, that the problem report or change request be reviewed and approved by a change control organization, and that the ensuing changes be managed and tracked by that problem report. In contrast, changing an HC2 item may simply require agreement from a technical lead or peer review. Whatever process and criteria will be used should be documented in the configuration management plan for the project.

The second component of change control—records, approvals, and traceability—applies only to HC1 items, and is intended to address the integrity of a change once the first component (integrity and identification) ensures that the change has been identified and authorized. HC1 items require a higher level of integrity during the change process, and this activity provides that integrity by requiring that changes be minutely managed throughout the evolution of the change.

DO-254 provides the following objectives for this activity:

- Changes should be recorded, approved, and tracked.
- Changes should be traced to the reason for the change.
- The impact of the change should be assessed to identify any ripple effects.
- Feedback is provided to affected processes.

This guidance is both explicit and ambiguous, in that it states fairly clearly what must be done to manage changes, but on the other hand it does not provide guidance on the level of integrity required for each activity and objective. While the level of integrity will certainly depend to some degree on the DAL of the hardware, even that guidance provides little tangible direction and leaves almost everything to the discretion of the reader.

Given that most HC1 items can affect the integrity of the configuration item, it makes sense that change control activities should have a high level of integrity regardless of the DAL.

Recording changes simply means that each change to a configuration item should be documented. Determining how well the changes should be documented can be determined by looking at the other configuration management activities, since this activity must be consistent with all of the other configuration management activities and their goals. One configuration management activity that is closely related to change control is baseline traceability. If the goals of baseline traceability and change recording are taken together, it follows that a suitable level of integrity for change recording is to document changes with enough detail to enable adjacent baselines to be recreated from each other.

Change approval is similar to, but different from, the authority to make changes that is part of the integrity and identification aspects of change control discussed above. The approval to make a change means a specific change has been approved and that someone has the authority to make the approved change. In contrast, the authority to make changes allows someone to change an item but does not specify what changes will be made. Nor does it guarantee that the changes that will be made are approved or desired. This is one important aspect of change control that provides greater integrity for the changes made to HC1 items.

So for change control for HC1 items, changes must be approved as opposed to just authorized. In practical terms, this means that all changes to an HC1 item must be approved before being made.

Tracking changes means managing and monitoring changes to an item throughout the evolution of the change, from initial inception to final disposition and verification of the change. Tracking can be accomplished through procedural means such as regular status meetings on changes and updates, or through the use of tools such as a problem reporting system. The goal of this activity is to make sure that all changes that are approved are implemented in the accepted timeframe and in the intended manner.

Tracing a change to the reason for the change is another tool in assuring the integrity of changes. It justifies every change that is made to ensure that they all have a valid reason and none are made randomly or arbitrarily.

Using problem reports for change control is highly recommended, if not implicitly required through the expectation of certification authorities. As stated previously, problem reports are ideally suited to recording, approving, tracking, and tracing changes, especially if they are used in the manner described earlier in this chapter. If the problem report guidance in this chapter is followed, and problem reports are created for all changes to HC1 items, then the recording, approving, tracking, and tracing aspects of change control never have to be addressed as a separate activity.

This interdependence between problem reporting and change control is not required by DO-254; the only reference to this tie-in is a note stating that there is a relationship between them because some changes may result from an issue that is managed by a problem report. However, using problem reports as part of the change approval process is a convenient means of satisfying the needs of both aspects of change control.

RELEASE

Release is the act or process of formally placing an item under configuration control and making it available for use by downstream processes and activities. It ensures that only authorized data is used, particularly for verification and manufacturing. Releasing data normally requires authorization, and once released, the item is typically stored in a configuration management system that allows the item to be retrieved when needed, but does not allow the item to be changed without the proper authorizations and approvals. Data items used for manufacturing must be released prior to use.

Releasing an item creates a baseline of the item, but it differs from the baseline activity in that it is a formal baseline and requires approvals, and the version that is released is placed under configuration/change control. Baselines can be created outside of configuration control and can be informal. For example, requirements can be baselined in a requirements management system, and an HDL design can be baselined in a configuration management system, but in both cases the baselines are informal until they go through a formal release process. There can also be multiple baselines of an item between releases, say for intermediate updates and peer reviews that may occur when a released item is being changed.

RETRIEVAL

All configuration items must be retrievable from their archive location. This should be self-evident, since an archive is of no value if the information in it cannot be accessed, but at the same time there are no guidelines on how accessible the data has to be, and it is possible to envision an archive where the data can be retrieved but not without great effort and extended delays.

The accessibility of archived data takes on added significance when the data must be retrieved on short notice and under stressful conditions, such as when a piece of airborne equipment experiences a failure that is serious enough to ground a fleet of aircraft well after that aircraft has entered service. When such an event occurs, the entire project may need to be reconstituted very quickly to fix the problem, prove that the problem is fixed, and get the fixed hardware back in production. Any delays in retrieving data and reconstituting the project will translate into additional time that the fleet will be grounded. The financial liability that can be incurred from such an event can be enormous, so the time it takes to retrieve archived data can have a significant impact under the right conditions.

The type of data that is archived can also have an impact on the ease or timeliness of retrieval. For example, a development environment, including test stands that are used for both design and verification activities, can be archived as fully assembled and operational hardware or as a set of drawings from which those test stands are built. Archiving test stands can be expensive, space-intensive, and may require periodic calibration or maintenance, whereas archiving the drawings will be much easier and convenient. However, under the extreme conditions described in the aircraft fleet grounding scenario described above, the time needed to reconstruct the test stands from drawings could add many weeks to the recovery schedule, and therefore increase by many weeks the time that the fleet is grounded. If any of the original components of the test stands are obsolete—a likely scenario considering the rapid evolution of electronic hardware and computers—the delays could be much greater because it will necessitate that the test stands be redesigned as well as rebuilt. Either approach will satisfy the DO-254 objectives of archive and retrieval, but one comes with a significantly higher risk in the long run.

The data describing the configuration item being retrieved must also have adequate integrity, meaning that after retrieval the data must be identical to the data that was archived. So if data was compressed for archive, the archive and retrieval process must ensure that the decompression format will still be supported after an extended period of time, or that the decompression software is archived with the compressed data. The same concept applies to data that is archived off-site at a data storage facility, especially if that facility is owned by a different company: precautions should be taken to ensure that the data will be preserved and retrievable if the storage company should suffer damage or go out of business.

Guidelines and procedures for retrieving configuration items should be established for each method used to store and archive the configuration item. Local storage and retrieval on a workstation or personal computer is maintained by the user. Network servers can have access rights and permissions established for each account and for the directories accessed by that account. An electronic file management system will have user login credentials and permissions. In general, an electronic file management system retains a copy of all data that has been checked in and forces the version number to change when a new version is checked in. Production data systems typically only allow general users to see read-only versions of the data for a configuration item. The production data system will allow users to access a PDF copy or other formats that cannot be modified. Retrieving data from an off-site long-term storage or archive facility should use more formal requests for a configuration item, especially for older items that may still be in paper or a hard copy format. Retrieving data from a backup is typically handled in conjunction with an information services department and requires a documented request.

DATA RETENTION

The Data Retention activity addresses the need to ensure that archived data is preserved intact for as long as the data is needed (typically as long as the hardware item is in service). It is not uncommon for modern aircraft to stay in service for more than 50 years, and there are still relatively ancient aircraft (such as the Douglas DC-3) still in commercial service throughout the world more than 75 years after they were built. If a piece of equipment on such an aircraft manages to survive without modification for the life of the aircraft, then the equipment manufacturer needs to archive that equipment's life cycle data for generations after the equipment designers have retired or died. While this scenario may be uncommon for most pieces of electronic hardware, it is still probable enough to make planning for it a wise idea.

Data retention requirements can also be traced back to the regulations. Federal Aviation Regulation 14 CFR 21.49 states that the holder of a type certificate must make the certificate available for examination upon the request of the FAA or the National Transportation Safety Board (NTSB). A type certificate is defined in 14 CFR 21.41 as including the type design, the operating limitations, the certificate data sheet, the applicable regulations with which the FAA records compliance, and any other conditions or limitations prescribed for the product. 14 CFR 21.31 defines the type design to consist of:

- The drawings and specifications, and a listing of those drawings and specifications, necessary to define the configuration and the design features of the product shown to comply with the requirements applicable to the product;
- Information on dimensions, materials, and processes necessary to define the structural strength of the product;
- The Airworthiness Limitations section of the Instructions for Continued Airworthiness as required by parts 23, 25, 26, 27, 29, 31, 33, and 35 of the regulations, or as otherwise required by the FAA; and as specified in the applicable airworthiness criteria for special classes of aircraft defined in 14 CFR 21.17(b); and
- For primary category aircraft, if desired, a special inspection and preventive maintenance program designed to be accomplished by an appropriately rated and trained pilot-owner.
- Any other data necessary to allow, by comparison, the determination of the airworthiness, noise characteristics, fuel venting, and exhaust emissions (where applicable) of later products of the same type.

In summary the type design consists of the data (design assurance data in this case), the type certificate includes the type design, and the type certificate must be made available upon request. In other words, the design assurance data must be made available upon request.

PROTECTION AGAINST UNAUTHORIZED CHANGES

Protection against unauthorized changes may sound familiar, since this topic was already introduced in the discussion on the first component of change control, in which the integrity of data should be preserved through protection against unauthorized changes. However, in this incarnation the protection against unauthorized changes is taken within the context of archived data, whereas in the previous instance it was discussed within the context of current active data. While the concept is the same—protect the data from unauthorized changes to preserve its integrity—the setting and processes may be a little different.

Like the change control version of this topic, there is no formal guidance on the types and levels of control that must be implemented; the final determination of the adequacy of the protection, as well as how long the protection should be maintained, will be made by the certification authorities. However, the integrity of the data is just as important in an archive as it is with current data, so it makes sense to implement similar controls.

MEDIA SELECTION, REFRESHING, DUPLICATION

Protection against unauthorized changes is just one aspect of ensuring the integrity of archived data; it will prevent data from being changed when it should not be, but it will not ensure the integrity of the archive itself. It is the Media Selection, Refreshing, and Duplication activity that addresses the integrity of the archive and therefore its contents.

Media selection is the process of selecting the method of archiving data, such as using a commercial data storage facility, an in-house networked server, or optical storage disks. Each medium has its advantages and disadvantages, and the archive process should consider all of them when selecting which medium to use. The selected medium should satisfy all of the objectives of the configuration management processes as well as the airworthiness requirements. Of particular interest is the issue of data retention, since it is conceivable that the data for some hardware items could require archiving for decades.

Media refreshing addresses the issue of data retention when the selected archive media has insufficient storage life to satisfy the long-term data retention requirements for an archive. If a medium that does not have long-term stability is selected, the archive process should anticipate the need to refresh the archive to mitigate the shorter than needed storage life.

Media duplication ensures that the archived data will be safe from loss due to destruction or deterioration of the archive. It is similar in concept to redundancy in electronic hardware design. Duplication or duplicate archives should be independent as well to prevent a single-mode failure, or in other words, more than one copy of the archive should be independently maintained to prevent a single event from causing a loss of the archived data.

Consideration also needs to be given to migrating data from one media to a subsequent or next generation media. Years ago, paper drawings and documents were stored. Microfilm was next used to reduce storage space. These days, electronic media is used to store data. The media has evolved from floppy disk to tape to CD-ROM to BluRay-ROM to Flash drives, and so on. Proper management of archived data over the ensuing years will allow the data to be accessed when needed.

REFERENCES

1. Code of Federal Regulations, Title 14: Aeronautics and Space, PART 21—CERTIFICATION PROCEDURES FOR PRODUCTS AND PARTS, Subpart B—Type Certificates, 21.31 Type design.
2. Code of Federal Regulations, Title 14: Aeronautics and Space, PART 21—CERTIFICATION PROCEDURES FOR PRODUCTS AND PARTS, Subpart B—Type Certificates, 21.41 Type certificate.
3. Code of Federal Regulations, Title 14: Aeronautics and Space, PART 21—CERTIFICATION PROCEDURES FOR PRODUCTS AND PARTS, Subpart B—Type Certificates, 21.49 Availability.

10 Additional Considerations

PREVIOUSLY DEVELOPED HARDWARE

Previously developed hardware, like the other topics in chapter 11, may or may not apply to a project depending upon the selected certification strategy.

PDH is invoked if there is an intention to reuse hardware that was developed at some earlier time. There are numerous reasons for reusing previously developed hardware, but for the purposes of this book the most relevant reason is to simplify and shorten the path (and cost) to DO-254 compliance and the eventual approval of the hardware.

Most PDH will come from one or more of the following sources:

- Commercial off-the-shelf (COTS) hardware or component
- Airborne hardware developed to other standards (e.g., a military or company standard)
- Airborne hardware that predates DO-254
- Airborne hardware previously developed at the lower design assurance level (DAL)
- Airborne hardware previously developed for a different aircraft
- Airborne hardware previously developed and approved, and then subsequently changed

The PDH itself can be from any level of hardware ranging from entire systems down to fragments of HDL code. Most PDH items will be one of the following:

- An entire system
- An LRU or box (including SW)
- An LRU or box (hardware only)
- An entire circuit card assembly (including SW)
- An entire circuit card assembly (hardware only)
- Part of a circuit card assembly (CCA)
- An entire PLD (device and HDL)
- An entire HDL design
- Part of an HDL design

Any of these items may have to be modified, retargeted, upgraded to a higher DAL, or used in a new way when used as PDH. In addition, modifying PDH may require the use of new (or newer) design tools.

Some common scenarios for PDH reuse are:

- A hardware item (LRU/box, CCA, or PLD) that is being reused without changes.
- A hardware item (LRU/box, CCA, or PLD) that is being modified for a new use.
- A hardware item (LRU/box, CCA, or PLD) that is being updated but will remain in its original system and application.
- An HDL design that is being retargeted to a new PLD device.
- Increasing the DAL for any of the above scenarios.

Section 11.1 of DO-254 defines four scenarios for PDH, all discussed in sections 11.1.1 through 11.1.4: modifications to previously developed hardware (11.1.1), change of aircraft installation (11.1.2), change of application or design environment (11.1.3), and upgrading a design baseline (11.1.4). While each of these sections address different aspects of PDH that may be applied alone, it is more common for more than one of them to apply to any given use of PDH. For example, if a PLD was developed several years previously for aircraft model X, using it on aircraft model Y would be a change of aircraft installation (section 11.1.2) that could require that the PLD be modified for its new installation (section 11.1.1), use a later version of design tools to make the changes (section 11.1.3), be upgraded to a higher DAL (section 11.1.4), and interface to a different circuit card assembly (section 11.1.3).

A key point in reusing previously developed hardware is to maximize the reuse of the approved compliance data from the previous development program. The amount of data that can be reused will depend on the level or type of hardware being reused: the more complete the reused hardware, the more data can be reused.

Reusing an LRU as-is in a new installation is optimal since all of the compliance data from the previous development program can be reused. If the LRU had to be modified for use in the new installation, the amount of reusable data would go down while the amount of new development and verification effort would go up.

For any use of PDH the specific data items and activities that have to be augmented or redone must be determined on a case-by-case basis. It is impractical to provide comprehensive guidance that can apply to all potential uses of PDH, and this is one of the reasons that the guidance in DO-254 does not attempt to provide any more detail than it does. However, it is possible to provide some additional summary guidance that may assist the reader in understanding what DO-254 alludes to when it discusses the use of PDH.

The strategy for the use of PDH should address the hardware item itself, its data, and its interfaces to its parent hardware item. All AEH items are ultimately components within a higher level piece of hardware (commonly called its *parent hardware*): HDL code is a component in a PLD device; a PLD or other electronic component is itself a component in a CCA; a CCA is a component within an LRU or box; and an LRU or box is a component in its overlying system or the aircraft itself. As such, when any AEH item is used as PDH, its performance and functionality must be addressed as a singular item and also with respect to its context as a component in its parent hardware item. Thus the PDH item's design data (requirements and design) must be complete in expressing the PDH item, but must also be integrated into its parent hardware by requirements traceability. Verification must confirm not only that the PDH functions as intended, but that it fulfills all functionality allocated to it from the parent hardware, and that it functions correctly when installed in the parent hardware.

When determining what must be done to the PDH and its data, the existing hardware and data must be assessed against the PDH scenarios defined in sections 11.1.1 to 11.1.4 of DO-254, along with the specific needs of the parent system.

Table 10.1 summarizes the data that can typically be reused, along with any additional effort that may have to be conducted, for some common uses of previously developed hardware. The information in Table 10.1 covers just a handful of common possibilities and is not intended to be comprehensive, nor does it provide guidance on how any use of PDH should or must be conducted. The data and activities in Table 10.1 are instead just intended to provide some additional insight into the PDH activities that DO-254 describes.

The intended strategy for PDH must be stated and described in the PHAC. An analysis should be conducted to identify any gaps between the existing hardware and data and what the new use requires. Any gaps that are identified should be filled with new data, which may include data from service history, additional verification, or reverse engineering. A change impact analysis should be performed; it should consider the effect on all aspects of DO-254 and the hardware life cycle, which may include planning, requirements, design, validation, verification, implementation, transition to production, configuration management, and tool qualification. The change impact analysis may need to be included with the PHAC.

Any changes to PDH must be assessed for their effect on the system safety assessment, as should the use of PDH in a new application or installation. If the system safety assessment indicates an increase in DAL for the PDH, the PDH and its data must be assessed against the requirements for the new DAL. The higher DAL may require additional changes to all aspects of the PDH's compliance to DO-254, which again may include planning, requirements, design, validation, verification, implementation, transition to production, configuration management, and tool qualification.

Service history can be used (but is not required) to support the use of PDH.

When discussing PDH with the FAA, be sure to consider that FAA definition of PDH from Order 8110.105. FAA Order 8110.105 states that PDH is airborne electronic hardware (simple or complex PLDs) that was approved before Advisory Circular AC 20-152 (June 2005). PDH also includes projects for simple or complex PLDs where the hardware management plans were approved before Advisory Circular AC 20-152 (June 2005). The FAA uses the term "legacy" systems for systems approved before AC 20-152.

COMMERCIAL OFF-THE-SHELF COMPONENTS USAGE

The majority of airborne electronic hardware is composed of commercial off-the-shelf parts ranging from the simplest passive components to the most complex integrated circuits. In fact, very little of most systems can be considered custom built. The concept of COTS also extends to assemblages of these components, such as subassemblies, programmed PLDs, commercially obtained intellectual property (IP) and library functions for PLDs, and even entire systems. However, while each component

TABLE 10.1
Common Uses of PDH and Typical Levels of Reuse

PDH Activity	Requirements	Design Data	PDH HW	Verification & Validation Data	Verification & Validation Activities
Reuse unmodified PDH in new parent HW	Reuse Create traceability to new parent HW	Reuse	Reuse	Reuse	Verify interfaces to new parent hardware
Modify PDH for use in new parent HW	Update Create traceability to new parent HW	Update	Update	Update for modifications	Verify new functionality in PDH Verify interfaces to parent hardware
Modify PDH for use in same parent HW (update PDH only)	Update Update traceability to new parent HW as needed	Update	Update	Update for modifications	Verify new functionality in PDH Verify new and related functionality in parent hardware
Upgrade DAL	Upgrade as needed to meet new DAL Upgrade traceability as needed to meet new DAL	Upgrade as needed to meet new DAL Upgrade traceability as needed to meet new DAL Address tool qualification for new DAL	Update as needed to meet new DAL	Upgrade as needed to meet new DAL Upgrade traceability as needed to meet new DAL	Perform as needed to meet new DAL Upgrade traceability as needed to meet new DAL Address tool qualification for new DAL
Change of design environment	Reuse	Reuse Address tool qualification	May require new baseline for HW	Reuse	May require reverification of PDH May require reverification of interfaces to parent HW Address tool qualification
Unmodified PDH interfaces to new SW	Reuse	Reuse	Reuse	Reuse	Verify all interfaces to new SW

in a system may be COTS components, DO-254 section 11.2 states that the certification process does not address individual components, modules, or sub-assemblies because those COTS components are addressed when the function they belong to is verified. Or in other words, if the AEH item is verified, then its constituent components are assumed to be verified as well. Thus for the most part individual COTS components do not have to be specifically addressed.

However, DO-254 also states that the basis for using COTS components is the use of an electronic component management plan (ECMP) in conjunction with the design process. Or in other words, using an ECMP is essential for establishing the pedigree and authenticity of all of the COTS components that are used in the AEH, thus providing supporting substantiation for the assumption that verifying the function or system will also verify the veracity of the COTS components.

The ECMP should satisfy items 1 through 7 of DO-254 section 11.2.1, which can be considered the ECMP's objectives or goals. These seven goals can be paraphrased as follows:

1. Ensure that all components were manufactured by companies that have a proven track record of producing high quality parts, and that none of the components are of questionable origin. For example, counterfeit components are often virtually indistinguishable from authentic parts but will often exhibit lower reliability or performance, so the ECMP should ensure that components are only purchased from original manufacturers through reliable sources.
2. As part of their overall reputation for excellence, the component manufacturers have a high integrity quality control program to ensure that all components are of consistently high quality and will always meet their rated specifications.
3. Each component that was selected for the AEH has successfully established its quality and reliability through actual service experience. In other words, the ECMP should ensure that new and novel components not be used until they have established their reliability in actual use.
4. Each component has been qualified by the manufacturer to establish its reliability, or else the AEH manufacturer has conducted additional testing on the component to establish its reliability after procurement.
5. The quality level of the components is controlled by the manufacturer, or if this cannot be established, additional testing is conducted by the component manufacturer or the AEH manufacturer to ensure that the components have adequate quality. In other words, the manufacturer should test all components to positively establish that every one will meet or exceed its specifications, and if they do not then the AEH manufacturer should.
6. Each component in the AEH has been selected for its ability to meet or exceed the requirements of its intended function in the AEH, including environmental, electrical, and performance parameters, or were screened through additional testing to ensure that all components will meet or exceed the needs of the AEH. So while it is best to use only components that are rated by the manufacturer to be in excess of what the AEH needs, components with lesser specifications can also be tested to identify individual components that exceed their specifications to where they meet the needs of the AEH.

7. All components are continuously monitored to quickly identify failures or other anomalous behavior, and if any are encountered, the deficiencies will be fed back to the component manufacturer to effect corrective action. In other words, keep track of component performance and pay special attention to failures or other unwanted behavior so components that are deficient or do not meet their rated specifications can be quickly identified and corrected by the manufacturer.

In addition to the component management process, DO-254 section 11.2.2 discusses areas of potential concern with respect to procurement issues. Procurement issues are not limited to simple availability and obsolescence, but include such issues as variations in quality between production runs, manufacturing improvements that may affect component performance in a way that can undermine design assurance, and the availability of design assurance data for COTS components that are considered complex or otherwise require compliance to the guidance in DO-254.

Obsolescence is one of the more serious and prevalent issues with regard to component procurement. When a component manufacturer discontinues a component used in an AEH item, the ripple effects can be challenging, expensive, time-consuming, and even affect the design assurance of the system. Reputable manufacturers will generally provide enough advance notice to allow customers to plan for component obsolescence, but even with plenty of notice the effects can still be significant. There is little official guidance on how it should be managed, but common sense indicates that the ECMP should (in addition to items 1 through 7 in DO-254 section 11.2.1) be vigilant for obsolescence and query manufacturers for long-term availability of the component both before it is selected and after the AEH has entered production.

Integrated circuits are well known for exhibiting measurable variations in performance depending not only on when they were manufactured, but even according to the location of each die on the silicon wafer. However, manufacturers should sort each component die by its performance to ensure that any part that is labeled with a specific part number will still meet the guaranteed performance specified in the component data sheet. Thus while variations in performance is listed as a concern for component management, mitigating this phenomenon may not precisely be a components management issue, but more a design issue in which designers should ensure that the electronic design in which the component is used will work properly regardless of where the individual components fall in the performance ranges specified in the component data sheet.

It is also common for integrated circuits with the same part number but different manufacture dates to show significant variations in performance, often toward improvements in performance. A common cause of such a variation is the shrinking of the feature sizes in the semiconductor device, which can often result in improved speed and performance. It is also common, however, for the improvements in performance to result in devices that perform too well for the design and cause instability or other anomalous behavior that does not exist when using the original devices.

Some of the more complex COTS components, such as assemblies or preprogrammed PLDs, may raise concerns with certification authorities and/or result in inadequate design assurance data if the manufacturer does not or will not provide the data with the hardware. AEH designers should consider this possibility when using such COTS devices in their designs. The same issue applies to COTS IP cores that are popular with HDL-based digital components. IP cores can be an attractive way to buy pre-designed complex functionality for PLDs, but when obtaining IP cores the availability or generation of design assurance data should be carefully considered before committing to their use. If the design assurance data is not available for whatever reason, it may have to be reverse engineered to meet the design assurance goals of DO-254 and other certification guidance.

DO-254 does not provide complete guidance on the use of COTS components. If further guidance is needed on preparing an ECMP, the International Electrotechnical Commission document IEC TS 62239, *Process management for avionics—Preparation of an electronic components management plan,* can be used as a guide.

PRODUCT SERVICE EXPERIENCE

Product service experience takes advantage of time that a component or AEH hardware has accumulated in actual operation by documenting the time in service and using that data to substantiate a claim that the hardware is safe and satisfies the objectives of DO-254. Service experience can be used to supplement or even fully satisfy the objectives of DO-254 for COTS devices or previously developed hardware. In both cases the use of service experience is optional; it can be used if it is available, but is not required to satisfy the objectives of DO-254. If service experience will be used, its use must be coordinated with the certification authorities and stated in the PHAC.

The use of service experience requires that current and/or previous use of the hardware be documented. How much time must be documented is not entirely standardized, and like many other aspects of certification it is evaluated per case by the certification authorities, so it is important to communicate early and often with the certification authorities about its use. The service experience does not have to be from an aerospace application, but logically the closer the service experience is to the intended application, the more relevant that experience will be and the more confidence that experience will inspire. The relevance of the service experience is one of the factors that will be assessed when determining its acceptability.

The relevance and acceptability of the service experience will depend on four criteria:

1. The relevance of the service experience to the intended use, as defined by its application, function, operating environment, and design assurance level. The more similar the service experience is to the intended use, the more relevant it is, and therefore the more acceptable the data is likely to be.
2. Whether the hardware that accumulated the service experience is the same version and configuration that is proposed for the intended application. If they are different, then it may be more difficult to justify the use of the service experience, or additional analysis may be needed to justify it.

3. Whether there were design errors that surfaced during the period of service history, and if so, whether those errors were satisfactorily dispositioned. The disposition of each error can be effected through elimination of the error, mitigation of the error's effects, or through an analysis that shows that the error has no safety impact.
4. The failure rate of the item during the service history.

These criteria are satisfied through four activities that are used to assess service experience data, or in other words, the following four activities should be conducted to determine whether the service experience adequately meets the four criteria:

1. Conduct an engineering analysis of the service history with respect to the item's application, installation, and environment to assess how relevant the service history is to the intended use. This analysis may look at a wide variety of data, including specifications, data sheets, application notes, service bulletins, user correspondence, and errata notices, to determine how relevant the service history is.
2. Evaluate the intended use of the item for its effect on the safety assessment process for the new application. If there were any design errors discovered during the service history, this evaluation should include approaches that can be employed to mitigate the effects of the errors, if applicable.
3. If there were design errors, the statistical data for the errors should be evaluated for their effect on the safety assessment process. If there are no statistics, the errors can be qualitatively assessed for their impact.
4. Evaluate the problem reports for the item during the service experience to identify all errors that were discovered and how those errors were dispositioned. While correcting any errors during that period is generally preferred, it is not required to enable the service experience to be used as long as the remaining errors are mitigated and/or shown to not affect the design assurance of the item. The mitigation can be implemented through architectural means in the new application, or through additional verification to show that the error will not be an issue in the new application.

Once the service history data has been evaluated and deemed to be sufficient to establish or supplement the design assurance for the item for its intended use, the data and its assessment should be documented to substantiate the claim for design assurance. The service experience assessment data should include the following:

1. Identify the item and its intended function in the new application, including its DAL. If the item is a component in a Level A or B function, there should be a description of how the additional design assurance strategies in DO-254 Appendix A will be satisfied, such as the use of architectural mitigation and additional or advanced verification, to establish the requisite design assurance.
2. The process used to collect and evaluate the service experience data, and the criteria that were used to assess whether that data was adequate and valid, should be described.

3. The service experience data should be documented. This data should include the service history data that was evaluated, any applicable change history, assumptions that were used in the analysis of the data, and a summary of the analysis results.
4. A justification for why the service history is adequate for establishing or supplementing (as applicable) the design assurance for the item in its new intended use and DAL.

In practical terms, service history should be based on the same component, with the same part number and version. The service hours should be commensurate with the failure probability required by the design assurance level. PLDs with millions of hours of service experience are suitable to propose for design assurance level A and B applications. Service experience with thousands of hours will not yield much credit for DO-254 compliance. Service experience from flight test programs are not suitable for demonstrating compliance to DO-254 since the aircraft is not yet certified.

TOOL ASSESSMENT AND QUALIFICATION

Tool assessment and qualification is one of the more misunderstood aspects of DO-254. It is not uncommon for the uninitiated to assume that the design tools used to produce Level A DO-254 compliant hardware have to be "special." The news that no special tools are required is generally greeted with relief, but on the other hand the news that other things must be done to avoid the use of "special" tools is greeted with a resurgence of concern.

The fundamental tenet of DO-254's approach to tools (in general, "tools" refers to computer-based design and verification software, or to electronic measurement tools used in an electronics lab, not to mechanical hand tools found in a shop) is that no tool can be trusted unless it has been proven in some way to always produce the correct output when given the correct inputs. There are a number of ways to satisfy this need to prove the correctness of a tool's output, and fortunately most of them do not require that the tool be "special."

There are two basic approaches to proving that a tool produces the correct output: the first approach is to comprehensively test and analyze a tool before it is used to prove that the tool will generate the correct outputs under the conditions under which it will be used, otherwise known as qualifying the tool; the second approach is to first use the tool and then test its outputs to independently prove that the outputs it produced were correct for the inputs it was given, otherwise known as verification. Both approaches will produce the same result in the long run, but have different considerations for their use.

Qualifying a design tool requires that the confidence in the tool be commensurate with the design assurance level of the hardware it is producing. In general, this means that the assurance for the output of the tool is at least as high as the end-item hardware, so for Level A hardware the design tool would have to have the same level of integrity as the Level A hardware that it generates. For most hardware design tools this would be a monumental task, especially for PLD design tools such as synthesis tools given their immense complexity and known propensity for introducing failure modes by altering the logic in the HDL (see the chapter on Design Assurance Through Design Practice).

The advantage of qualifying a tool is that the tool output can be trusted and is thus exempt from further scrutiny. For a design tool it means that the output of the tool will not have to be verified because the qualification process essentially verifies the tool. Qualifying (i.e., verifying) the tool or verifying the tool's output both require significant effort and time; the path that is taken depends on the needs of the program and should be preceded by a thorough analysis of the comparative benefits, both immediate and long term. Since tool qualification is perpetual—once qualified, the specific version and configuration of the tool that was qualified can be used from that point on and for multiple projects—future uses of the tool should be considered when assessing the long-term benefits of qualification.

The disadvantage of qualifying a tool is that for some tools the cost of qualification can be enormous, often exceeding the cost and time of verifying its outputs. For tools that will not be used beyond the current project, verification of its outputs can be the most efficient and least painful option. An additional consideration when weighing the cost versus benefits of tool qualification is that verifying the tool's output, if properly managed and documented, can be leveraged in the future as a means of satisfying some or all of the qualification criteria for the tool for future uses. Thus if it is known that the tool will be used for future projects, documenting the independent assessment of its outputs for one project can be used to support qualification of the tool for future projects or even future evolutions of the original project.

Tools are divided into two types: design tools, which are used to generate the hardware design; and verification tools, which are used to verify the design. There is also a verification tool sub-type commonly known as verification coverage tools, which are used with elemental analysis to assess the completion of verification testing, or in other words to measure the extent to which the verification process tested the individual elements of the design.

Design tool examples would be the synthesis and place and route tools used to convert an HDL design into the programming file for a PLD, or the schematic capture and circuit card layout tools used for circuit card designs. Verification tool examples would include HDL simulators used for conducting formal simulation verification on the source code and post-layout models for PLDs, or an automated test stand that is used for conducting hardware tests on an LRU, or a logic analyzer and oscilloscope used for open-box verification and testing. An example of a verification coverage tool would be a code coverage analysis tool that is used during simulation to measure how thoroughly each line of HDL source code was exercised during simulations.

Design tools are treated most stringently with respect to tool qualification because a design tool has the capability to introduce an error into the design, so a flaw in the tool has a high probability of causing an error in the design and thus a reduction in design assurance. Verification tools are treated more leniently because the worst they can do is fail to detect an error in the design; they cannot introduce an error into the design, and failing to detect a design error would require the alignment of the verification tool's flaw with a specific type of design error, so the likelihood that a flaw in the tool can reduce design assurance is much lower than for a design tool.

Coverage tools are treated most leniently—they require no assessment or qualification—because they cannot materially reduce the design assurance of the hardware, since all they do is evaluate how thoroughly the design has been covered by the verification process.

The essentials of tool qualification can be summarized as follows:

- A design tool does not have to be qualified if:
    - Its outputs are independently assessed, or
    - It is used for Level D or E design.
- A verification tool does not have to be qualified if:
    - Its outputs are independently assessed, or
    - It is used for Level C, D, or E verification, or
    - It is only used to measure verification completion.

Note that laboratory verification tools, such as logic analyzers, oscilloscopes, function generators, and the ubiquitous handheld meters, are generally exempt from formal qualification considerations as long as their calibration is up to date and they have been tested to confirm that they can correctly measure the signal types that they will be encountering. Calibration and testing, in conjunction with the tools' widespread use, will generally suffice as independent assessment of their outputs.

Automated test stands may or may not require qualification depending upon how much they do on their own. A test stand that simply measures and records input and output signals, which are then manually reviewed, can normally get by with limited testing to show that they correctly measure and record representative signals of the types that will be measured, as would be done for other laboratory verification tools. An automated test stand that also evaluates the signals it measures and independently determines the pass/fail result will have to be qualified.

Figure 11-1 in DO-254 is a flow chart for the tool assessment and qualification process. Each block in the flow chart has a corresponding paragraph in section 11.4.1 that describes the activities that should be conducted for its corresponding block. Most of the blocks and their text descriptions are easily understood and are more or less self-explanatory, but it is still worthwhile to discuss the activities associated with each stage of the process. All of the information from the tool assessment and qualification process should be documented in the tool assessment and qualification data, which can then be recorded in a report if desired.

The first step in the process is to identify the tool, meaning that the tool name, model, version, manufacturer, and host environment (type of computer and operating system) should be documented in the tool assessment and qualification data.

The second step is to identify the process that the tool supports, which is another way of saying that the specific application of the tool should be used to identify the tool as either a design tool or a verification tool. For example, a synthesis tool is used only during the synthesis stage of the PLD design process, and a place and route tool is used only in the layout stage of the PLD design process, so both will be classified as design tools. In some instances a tool may be used for both design and verification; in that situation, the tool should be addressed separately for each type of use. Part of this activity is to define how the tool will be used and not used, such as identifying the scope of use for the tool; for example, if a single tool is capable of both synthesizing HDL code and also placing and routing the design, but the intent is to use only the place and route function in conjunction with a third-party synthesis tool, it must be made clear which of these functions will be used. Likewise, if the tool has limited capabilities and will therefore only perform certain functions, those limitations should also be documented. Finally, the output that the tool will produce (examples: an HDL netlist from a synthesis tool, a programming file from a place and route tool, a schematic diagram from a schematic capture tool, a waveform from a simulation tool, a text log file from a simulator running self-checking testbenches, etc.) should be identified and documented. All of this information should be documented in the tool assessment and qualification data.

The information gathered in this step has a direct bearing on the activities that need to be conducted for the tool qualification process. Identifying the way the tool will be used will identify the tool as either a design or verification tool, which will affect how the tool must be addressed during the tool assessment and qualification process. Identifying the role and scope of the tool will identify the scope of any required assessment and qualification that occurs during the process. Identifying the tool's output will define the specific activities that must be conducted to either verify the tool's output or qualify the tool.

Once this information is documented, the third step asks whether the output of the tool will be independently assessed. For most projects, this is the most important step in the tool qualification process because qualifying a tool can be avoided by independently assessing the tool's output. For a design tool, assessing the tool's output can be accomplished through independent verification of the hardware. For verification tools, the assessment can be accomplished through a manual review of the outputs or by comparing the tool's output (also done manually) to a similar tool's output.

For PLD designs, the HDL text editor (it sometimes comes as a surprise to engineers that a text editor is considered a design tool, but as the primary means of creating an HDL design it definitely qualifies) output consists of HDL text files that are independently assessed during the HDL code review, so no further action on that tool is required. The synthesis tool generates an HDL netlist that describes the reduced and optimized logic, and when manually reviewed this netlist is essentially unintelligible, so assessing the output of a synthesis tool makes no real sense. Likewise, the programming file that is the output of the place and route tool has no human-readable content, and it has no meaning other than as the input to the PLD device programmer, so assessing its output is meaningless as well. There are design equivalency comparison tools that can compare the HDL to a netlist and verify that they express the same functionality, but if those tools are used as part of the formal assessment process they too will have to be subjected to the tool qualification process as well, so their use may not be cost-effective in the long run. The inability to independently assess the outputs of the synthesis tool and the place and route tool would seem to present an obstacle to assessing their outputs and relieving them of the need for qualification. However, DO-254 settles this quandary by stating that if independent verification conducted on the finished design indicates that the design is true to its intended functionality, then the tool chain that created the design can be assumed to be functioning correctly. So in effect all of the tools in the PLD design tool chain (the synthesis tool, place and route tool, and device programmer, and even the text editor as well) are simultaneously covered by the verification process even though the output of each tool is not assessed separately. In this situation the design equivalency comparison tool is not even needed, since the equivalency between the HDL and its synthesized netlist is actually irrelevant if the finished product (the programmed PLD) is verified and proven to be correct. Such a tool can be used informally to provide added confidence in the performance of the synthesis tool, and its use for that purpose should be encouraged, but as a formal step in the tool assessment and qualification process its use is unnecessary. When design equivalency comparison tools are used for rehosting designs or converting an FPGA into an ASIC, they should be assessed for that purpose.

The design and verification processes defined in DO-254 include all the necessary processes and activities to eliminate the need to qualify any design tool, so for almost all projects, qualifying a design tool will be an elective task that is driven by extra-project considerations such as the future use of the tool (as discussed previously).

Verification tools, on the other hand, are not so easily assessed. Verification and verification tools normally play a large part in eliminating the need to qualify a design tool, but while verification follows design and provides the assessment for design tools, there are no processes that follow verification that can provide the same service for verification tools. The assessment of verification tools must be conducted as part of the verification process, or be conducted as a separate activity that is independent of both the design and verification processes.

DO-254 provides two suggestions on how to assess the outputs of verification tools: manually review the tool outputs, or compare the output of the tool against the output of a comparable (but dissimilar) tool. For the first suggestion, that of manually reviewing the tool's outputs, an example would be to manually review the waveforms from a PLD simulation against waveforms captured during in-circuit hardware tests. For the second suggestion, that of comparing the tool's output to the equivalent output from a comparable but dissimilar tool, an example would be to run the same set of PLD simulations with two dissimilar simulators and compare the waveforms.

If the outputs of a tool are independently assessed as discussed in the previous paragraphs, no further action is necessary. The assessment of the tool, including the rationale and results of the assessment, should be documented as part of the tool assessment and qualification data.

If a tool's output is not independently assessed, the fourth step of the tool assessment and qualification process identifies the design assurance level of the function that the tool will support. If the tool is a Level D or E design tool, a Level C, D, or E verification tool, or if the tool is used to measure verification completion, then no assessment or qualification is needed. If the tool is a Level A, B, or C design tool, or a Level A or B verification tool, then additional assessment is necessary. The designation of the tool according to its function and DAL should be documented in the tool assessment and qualification data.

In the fifth step of the process, Level A, B, and C design tools, and Level A and B verification tools, whose outputs are not independently assessed, can avoid further assessment and qualification if they have significant relevant history. Relevant history is an attractive alternative to tool qualification, or in some cases and depending on the tool, can be used as part or all of a tool's qualification. However, at the same time the criteria for invoking relevant history is not clearly defined in DO-254. DO-254 states that no further assessment is necessary (or, given that any tool that has reached this part of the process has not been assessed, DO-254 is actually stating that no assessment is necessary) if it is possible to show that the tool has been used previously and produced acceptable results. This guidance is somewhat (but justifiably) ambiguous in that it does not include quantifiable criteria for what constitutes relevant history, which leaves the determination to the certification authorities. In turn, FAA Order 8110.105 CHG 1, Section 4-6 provides guidance to certification authorities on how relevant history should be justified, and while this guidance is more specific than DO-254, it still leaves the determination of acceptability to the discretion of the certification authorities.

According to DO-254 and FAA Order 8110.105 CHG1, relevant history, if it is going to be used, should meet the following criteria:

- The tool history can be from airborne or non-airborne use.
- The tool history must be documented with data that substantiates its relevance and credibility.
- The justification for the tool history should include a discussion of the relevance of the tool history to the proposed use of the tool. "Relevance" in this context refers to the way in which the tool was used, the DAL of the hardware it designed or verified, the type of data it produced or measured, and the specific functionality of the tool that was used.
- The tool history should prove that the tool generates the correct result.
- The use of tool history should be documented in the project PHAC.
- The use of tool history should be justified early in the project.

If relevant history is claimed, the tool assessment and qualification data should include a thorough discussion of the history, including how the history satisfies the criteria listed above. Any use of a tool can be used in the future for relevant history, so if it is possible that a tool will be used again for a similar application, the details of the current use should be carefully documented as if its use as relevant history has already been decided. Creating the documentation will take little time and could save a great deal of effort for future programs, or even for a subsequent evolution of the same program to implement a change in the design.

If relevant history is not applicable or will not be claimed, the tool must be subjected to the qualification process, starting with establishing a baseline for the tool and setting up a problem reporting system for the tool and its qualification, as described in step 6 of the tool assessment and qualification process. This means the specific version and configuration of the tool should be placed in a configuration management system and treated like any other piece of configuration managed data, including the assignment of an unambiguous configuration identity. This information should be documented in the tool assessment and qualification data.

Once the tool has been baselined, a "basic" tool qualification should be performed. This activity essentially tests the tool against its documented (as in a user's manual) performance and functionality. It requires that a basic tool qualification plan and procedure be generated and executed, using the tool's documented functionality and performance as requirements to be verified. This information should be documented in the tool assessment and qualification data. In addition, if the tool that is used for the actual design or verification activities differs from the tool that was baselined and then qualified, the tool assessment and qualification data must include a justification for using the different version and substantiating data for why it was acceptable to do so. For example, if an HDL simulator was qualified prior to the start of verification, and it was then discovered that the version that was qualified could not be used for formal simulations but that the next incremental version could, the use of the newer version without repeating the qualification must be justified. An example of a justification might be that the changes between the two versions fixed the problem that prevented the earlier version from being used but did not affect any other aspect of the tool's capabilities and functionality, including errata sheets from the tool manufacturer that pinpoint all of the changes that were made. The tool assessment and qualification data should include the basic tool qualification plan, the tool requirements that were verified and the test procedures that verified them, the qualification results, how independence was maintained during the qualification, and an interpretation of the results that support the qualification result.

Basic qualification constitutes the entire tool qualification effort for all verification tools and for Level C design tools, whereas Level A and B design tools must be subjected to a "full" tool qualification program. DO-254 provides no guidance on this type of tool qualification because of its variability and because it will be unique for each tool, providing instead some generic guidance to point the applicant in the general direction of how the process should be conducted. As starting points it suggests the use of the guidance in Appendix B of DO-254 and the tool qualification guidance of DO-178, plus any other means acceptable to certification authorities. So in practical terms, if an applicant wants to try qualifying a design tool, it is pretty much an open field where all terms, conditions, and requirements must be negotiated with the applicable certification authorities.

This type of tool qualification is very rarely attempted, and in most cases it is not a cost-effective route. However, if a design tool is to be qualified, it will require a highly formalized and structured qualification effort with its attendant plans, procedures, reports, analyses, traceability, problem reports, etc.—essentially the same as what would be conducted and documented for a formal verification effort. The rigor of the qualification process will be determined in part by the nature of the tool, and proportional to the design assurance level of the hardware. Any design tool qualification effort may also require significant participation from the tool manufacturer, including access to proprietary design information that the manufacturer may not be willing to provide. A tool accomplishment summary can be used to document the results of the tool qualification effort.

Both basic and full tool qualification need not verify the tool for all of its functionality, but can be limited in scope to just the functionality that will be used. For example, if a synthesis tool is going to be qualified and it will only be used to synthesize VHDL, then the Verilog functionality of the tool does not have to be qualified. Likewise, if a simulator can simulate both VHDL and Verilog, and only Verilog will be simulated, then the basic qualification (or the relevant tool history) need only cover its Verilog capabilities.

Overall, the most common approach to design tool qualification is to conduct formal verification instead of qualifying the tool, and the most common options for assessing the outputs of a verification simulation tool (for Level A and B hardware) are:

- Perform basic tool qualification
- Use relevant history if it exists
- Run the simulations on two dissimilar simulators, compare the results, and justify any discrepancies
- Use the same test cases (inputs and expected results) for both simulations and in-circuit hardware tests
    - Compare the results from simulations and electrical tests, and justify any discrepancies
- And as usual, other methods can be proposed in the PHAC for consideration by certification authorities

Table 10.2 provides an overview of the possible tool qualification outcomes according to tool type, DAL, independent output assessment, and acceptable relevant history.

TABLE 10.2
Tool Qualification Outcomes

Tool	DAL	Output Assessed	Relevant History	Qualification
Design	All	Yes	N/A	Not Required
Verification	All	Yes	N/A	Not Required
Design	A,B	No	No	Design Tool Qualification
Design	A,B	No	Yes	Not Required
Design	C	No	No	Basic Qualification
Design	C	No	Yes	Not Required
Design	D,E	N/A	N/A	Not Required
Verification	A,B	No	No	Basic Qualification
Verification	A,B	No	Yes	Not Required
Verification	C,D,E	N/A	N/A	Not Required
Verification Completion	N/A	N/A	N/A	Not Required

11 Summary

The old television cliché where the police detective tells the suspect that he can take the easy way by cooperating or the hard way by resisting, can apply to DO-254 compliance as well. Experience has shown that the easy way to DO-254 compliance is almost always through cooperation and an earnest desire to learn, adjust, and comply, as opposed to resisting and looking for loopholes in an effort to avoid compliance as much as possible. In fact, the seemingly eternal quest to avoid DO-254 is ironic in that it can take more time and effort to avoid DO-254 than it does to just embrace it and comply with it.

Taking the hard way or the easy way is a choice, and one that eventually confronts all airborne electronic hardware (AEH) developers. The choice that is eventually made will be based on considerations that are specific to the organization making the decision. Whatever the choice is, it will not be right or wrong—it will simply reflect the path that the organization wants to follow, and should only be judged according to how well it will work for them. Some choices will work better than others, and while choosing the easy way might be a better way, it is not the only way, nor is it the "right" way.

As discussed in the introductory chapter of this book, DO-254 is a compendium of the tried and true engineering best practices employed by the AEH industry, and in particular by companies that are known for producing the highest quality and safest aircraft in the world. As such it contains a wealth of useful information that can, in the long run, result in fewer problems, which in turn can translate into lower risk and even lower overall cost. The benefits of understanding and employing the processes in DO-254 can be especially valuable for companies that want to improve their engineering processes and culture, and/or that want to enter the AEH marketplace—those companies would do well to use DO-254 as a textbook on how to create a high integrity engineering culture. Trying to learn the equivalent skills on their own would be significantly more difficult and time consuming than simply embracing DO-254 and learning what it has to offer.

Much of DO-254 is fundamental engineering knowledge. For instance, the design process in Chapter 5 has been part of the engineer's basic tool box for generations. There is nothing new or mysterious about it, and it has proven itself over the generations to arguably be the best way to conduct a design effort, so while it may be common knowledge it is still one of the industry's best practices and is therefore rightfully included in DO-254. In contrast, DO-254 also contains other techniques, including most of Appendix B, that are more esoteric or even arcane, but they also have their roles in boosting the design assurance of a system or component, and so they too rightfully belong in DO-254. So in the final analysis there is little in DO-254 that has no role to play in supporting design assurance.

If DO-254 has any weaknesses, perhaps the most significant would be its reliance on requirements as the lynchpin for its design and verification processes. Or more realistically, the weakness is not in its reliance on requirements, because the use of requirements for both design and verification is one of the most effective and reliable means for expressing, implementing, and verifying functionality. The weakness only appears when the design and verification processes are based on poor quality requirements. Requirements are the input to the processes in DO-254, so the use of poor quality requirements will (not might) weaken and undermine the integrity of the output (the electronic hardware). Ironically, requirements are among the weakest and most poorly managed aspects of engineering in many engineering environments, and at the same time they are among the most difficult to improve. If ever there were a perfect example for bad habits dying hard, bad requirements practices would consistently be a prime contender for the title.

So if the lynchpin of the DO-254 design and verification processes is good requirements, then it is only logical to make sure that the requirements are of high quality. However, DO-254 does not provide detailed guidance on the characteristics of high quality requirements that are optimal for its processes, only that requirements should define the intended functionality of the electronic hardware. The transition from intended functionality to optimized requirements is what this book attempts to describe in its Requirements and Validation chapters. While those chapters provide an introduction to how requirements can work with DO-254 rather than work against it, they cannot provide the highly detailed treatment that will enable engineers to learn all that there is to learn about it. Giving requirements the full treatment to facilitate mastering those requirement skills would require a book of its own, and is beyond the scope of this book anyway: while the characteristics of requirements can be critical to the processes in DO-254, they are not really within the scope of DO-254 and are included in this book as additional guidance that can potentially save practitioners a great deal of money, time, and frustration. It can also facilitate the understanding of how the processes in DO-254 are intended to work.

If the attributes of a highly effective, efficient, and productive design program were listed, they would look similar to the contents of DO-254:

- Plan all aspects of the project to ensure that:
    - Risk is minimized.
    - The product is built right and delivered on time.
    - The customer is in agreement with what is being built.
    - The certification authorities are in agreement with the project and its product.
- Use a design process that will:
    - Use independent reviews to reveal as many errors as possible.
    - Systematically decompose functionality from the top down, and record the decomposed functionality in additional requirements and traceability.
    - Base the design on written requirements so everyone will be in agreement on what the electronic hardware has to do.

- Confirm that the requirements are actually correct.
- Catch errors as early as possible by confirming at multiple stages in the design process that the design is being developed properly.
- Compare the design to its requirements to ensure that the design satisfies all of its requirements.
- Use a verification process that will:
  - Confirm that the design is correct by comparing it against the same set of requirements that was used to create it.
  - Confirm that the design does what it is supposed to do, not what it was designed to do.
  - Test the design in a repeatable and documentable way.
  - Confirm that the design was completely verified.
- Control the design and its data to make sure that:
  - Every version of every data item is tracked and protected.
  - None of the data will be misidentified or lost.
  - No one can make unauthorized changes.
  - Every problem is documented and tracked to closure.
- Monitor the project to ensure that everyone is following the rules and not taking potentially damaging shortcuts.
- Where possible reuse previously designed electronic hardware to reduce cost and risk.
- Manage electronic components to minimize risk from obsolescence and unreliable suppliers.
- Make sure the tools always produce the correct output.

These attributes address the most common sources of error in an attempt to address as many error sources as possible.

Proper planning is essential for identifying and mitigating as many sources of risk as possible. Addressing technical and safety related risks (another way to describe design assurance) are especially important for AEH programs, and is really what DO-254 compliance is all about. It also provides transparency into the entire program for the aircraft manufacturer and the certification authorities so they will know that the AEH will be developed, tested, and manufactured in a way that will not only produce high integrity and reliable electronic hardware, but that it will comply in every way with the certification basis. Writing the plans and standards during the planning process is not about satisfying a DO-254 objective; the early transparency provided by the planning process is critical for establishing confidence between the AEH supplier, the aircraft manufacturer, and the certification authorities. The plans also establish an agreement between the three parties on both the tasks that will be performed and on the expectations of all three parties that can prevent any of them from wandering too far off course. So from the AEH supplier's perspective, the plans and standards should be approached as a contract that limits the customer or regulators from imposing additional design assurance related burdens, or from changing their minds about how the project will be conducted. In other words, if approached with the right perspective the planning process is the AEH supplier's friend and should not be put off as long as possible.

Engineers will sometimes object when first confronted with a structured design process, but most will eventually turn around once they have seen the benefits. The engineer stereotype seems at odds with this initial reluctance to embrace a design process because logical minds often seek the kind of structure and order that a design process offers. However, designing hardware can seem more like fun (as opposed to work) when conducted without the restraints of a design process. The problem is that designing AEH is a very serious business that can be fun under the right circumstances, rather than being a fun business that can create serious AEH designs under the right circumstances. The difference is where the emphasis is placed, and the correct emphasis is on the AEH design rather than on the fun of creating it. That does not mean that designing AEH cannot be a satisfying and fulfilling experience, and for most engineers it is, but it does mean that the safety of the AEH should take priority, and that priority should be the basis for the design process.

Like other aspects of DO-254 compliance, complying with a structured design process is a lifestyle choice that can make things easier or harder depending on the choice. In addition, the way in which the design process is complied with can make even more of a difference in how easy or hard the process becomes. This book includes guidance on how complying with DO-254 can be made as easy as possible:

- Plan ahead
- Perform the activities in the intended manner and order
- Combine activities (for example, performing requirements capture, functional decomposition, and traceability at the same time) to reduce needless rework and also to make all of the activities as effective as possible
- Invest heavily in writing requirements that will provide high integrity functional expression and harmonize as much as possible with the processes in DO-254
- Employ effective verification but still focus on creating a high quality design that will not need verification (in other words, do not use verification as a kind of safety net by skimping on the design under the assumption that if any errors are made the verification process will find them)
- Approach the processes in DO-254 not as separate activities but from the higher perspective where they become a unified web of complimentary interlocked and interacting processes
- Use a high integrity design philosophy that is focused on preventing potential errors (Design Assurance Through Design Practice)
- Always pay attention to configuration management

Verification is one of those processes that is sometimes treated as being somehow inferior to design engineering. It can be tempting to treat verification as a peripheral activity that is performed more because it has to be done than because it should, and to make it entirely separate from design. This separation between design and verification, while good from the independence perspective, can be damaging if carried too far. Independence does not require separation; in fact, close cooperation, communication, and synchronization between the design and verification activities are far more efficient and effective than building a wall between them.

Verification should be embraced as the partner of design, and it should be performed as thoroughly as possible. There can be a temptation to minimize verification because it can be expensive and time consuming even when optimized according to the guidance offered in this book, but when tempted to shortcut verification, consideration must be given to the fact that *every piece of hardware will always be completely verified one way or another*: what is not tested in the lab will eventually be tested in the aircraft. Since what is not verified in the lab will be verified in flight, the question of how much verification to perform should be answered only after full consideration of the potential effects of discovering an error in service, including the potential of a catastrophic failure. No verification program can find every possible failure mechanism, but programs often find their most interesting and potentially serious failure sources under the most obscure testing conditions (i.e., robustness verification). Ironically, it is the obscure testing conditions that are often seen as being the least productive and therefore are the most tempting to cast aside.

Configuration management is not the most glamorous or interesting process, but it plays an important role in ensuring the AEH is designed, tested, and produced faithfully and as free of errors as possible. Without it the potential for serious errors would be considerably higher. Likewise, process assurance may operate in the background, but its effect on design assurance can be profound.

Reusing previously developed hardware is a good business model and can result in higher quality hardware with a higher level of design assurance at less cost. This is ideal from the business, design assurance, and certification perspectives, so reuse should be employed where possible.

DO-254 has a total of thirty-four objectives that need to be satisfied to comply with the guidance it provides. One of the thirty-four objectives is to develop a design from the requirements. Consider the implication: one thirty-fourth or 2.94 percent of the objectives in DO-254 align with what engineers are trained, educated, and experienced at doing—designing electronic hardware. The other 33 objectives, or over ninety-seven percent of DO-254, entails planning, managing, standardizing, verifying, and validating the requirements and design and associated data. The overwhelming majority of objectives in DO-254 concern topics other than creating a design. It behooves us to learn, develop, and place value on the requisite skills for the other aspects of design assurance that are as refined as the engineering design skills themselves.

No process will eliminate all errors, nor can they completely plug all sources of errors, but employing countermeasures against the common error sources can still chip away at errors and noticeably improve the design assurance of the product, in turn ensuring that the AEH on our aircraft is as safe as it possibly can be. In the end, that is what DO-254 is all about.

Index

D

E

F

G

H

I

J

L

M

N

O

P

S

T

V

W